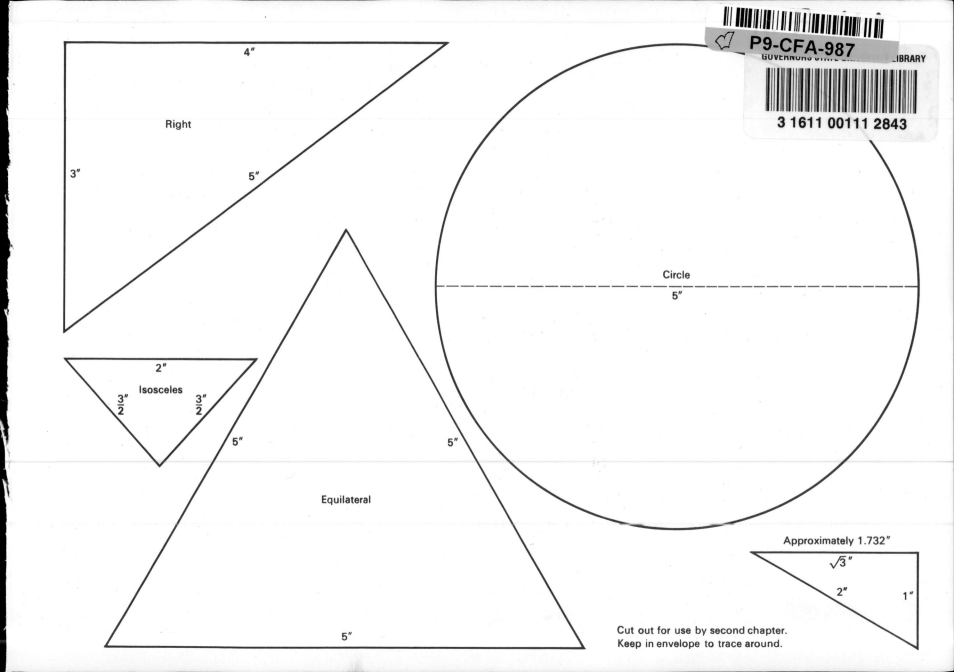

Right

4"

3"

5"

Circle

5"

2"

Isosceles

$\frac{3"}{2}$ $\frac{3"}{2}$

5"   5"

Equilateral

5"

Approximately 1.732"

$\sqrt{3}$"

2"   1"

Cut out for use by second chapter.
Keep in envelope to trace around.

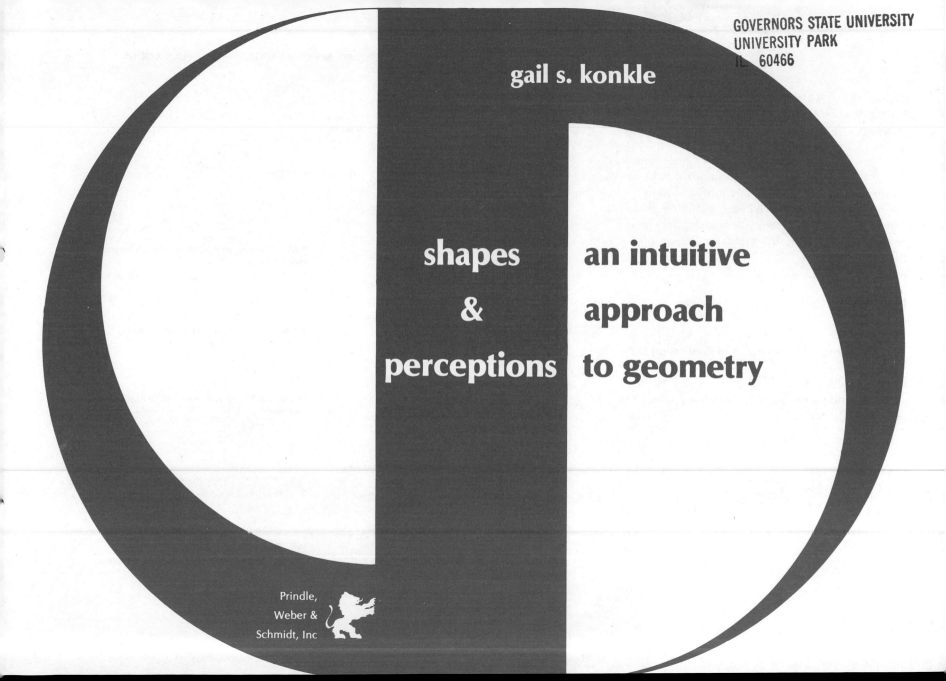

gail s. konkle

shapes
&
perceptions

an intuitive
approach
to geometry

Prindle,
Weber &
Schmidt, Inc

THE PRINDLE, WEBER & SCHMIDT SERIES IN MATHEMATICS EDUCATION

© Copyright 1974 by
Prindle, Weber & Schmidt, Incorporated,
20 Newbury Street, Boston, Massachusetts, 02116

Printed in the United States of America

Library of Congress Cataloging in Publication Data

Konkle, Gail S.   1937–
    Shapes and perceptions: an intuitive approach to
geometry.

    Includes bibliographies.
    1. Geometry. I. Title.
QA445.K64   516   74-2464
ISBN 0-87150-164-3

Consulting Editor:
John Wagner
Michigan State University

TITLES IN THE SERIES

preface

This text presents geometrical concepts and a few topological concepts intuitively at a level appropriate to the college student. The text is primarily a contribution in the realm of informal geometry. The intended audience includes those who want to understand and use elementary geometric concepts, those who need an introduction to coordinate geometry or formal geometry, and those who plan to teach geometry in the grade schools.

Actual prerequisites for the course involve knowledge of the shapes referred to by the words, "cube, square, rectangle triangle, sphere, and circle," plus mathematical sophistication commensurate with a previous course in arithmetic or algebra above the junior high level. For prospective elementary teachers, it seems wise to suggest that this course follow a course on the number system. The text embodies the content and spirit suggested by the Panel on Teacher Training of the CUPM in their call for one geometry course amongst four for the elementary teacher. This text also fulfills needs of the student whose instructor wishes to experiment with content other than the usual formal Euclidean geometry.

The major objectives of this text are (1) to provide the student with a more-than-intellectual understanding of basic geometrical concepts and contributing topological concepts, (2) to involve the student in that "discovery learning" which increasingly influences education, (3) to provide each student with some experience in guiding others and in explaining his own line of experimentation and reasoning, (4) to give the student a usable knowledge of key vocabulary, and (5) to expose prospective teachers to modern explanations and problems written for children.

The usage and terminology are chosen to be mathematically sound and to prepare the student for college level study of Euclidean geometry, projective geometry, topology, analytical geometry, and integration. No usages are misleading to future mathematical work. The text differentiates between what is defined intuitively as a conceptual aid for the student and what is defined mathematically.

The general organization of the text is governed by a desire to maintain mathematical correctness and continuity while taking advantage of the recent insights regarding psychology of learning. Ideas have been organized into chapters according to content and increasing level of abstractness and conciseness. The content of Chapters 1–15 proceeds from simpler to more intellectually-structured ideas, from topological equivalence to Euclidean congruence and similarity and measure problems. Chapters 16 and 17 are devoted to short immersions in coordinate geometry and formal geometry, while pointing out differences and relevances to the previous material and giving bibliographies of suitable material for further study, if that is desired. There is also an appendix explaining set intersection and union for those who face this text knowing only traditional mathematics.

The author finds here more than enough material for a 3 semester-hour course. A sophisticated and well-prepared class can assimilate chapters 2, 4, 7–11 quite rapidly (at the rate of one a day or better), allowing for leisurely explorations in other parts of the text. Here are a few suggestions of ways to use the text. (1) A class wishing to stress topology might spend more time on Chapters 1–6, whereas a class wishing to skip topology could skip all of Chapters 3, 5, and 6 except a few nontopological definitions. (2) A class emphasizing coordinate geometry or formal geometry might supplement with material from that suggested in the chapter bibliographies and skim topology or measure chapters,

The text directs itself to involving the student, cultivating his ability to estimate and think something through without reference to a memorized formula. Especially popular with the student are the informality of the text's approach and the introduction of many words and ideas in examples preceding formal definition. A chance to "play" with some thoughts and things often precedes an individual's discovery of relationships. For this reason, "play chapters," providing units of experience and exposure to new words, are placed strategically about every third chapter. They are meant to familiarize the student with concepts and vocabulary preceding presentation of related material, not to finalize anything.

The text helps the student by giving important pointers in a Preface to the Student. In that preface, the student is advised to read each section assigned without skipping an exercise when it occurs in the expository material. The instructor can help the student by assigning that preface and by reminding him of his responsibility to try each exercise as he comes to it in a discovery-oriented text such as this. Each chapter ends with answers to most exercises and problems in that chapter, for the author feels that it is important to the learning process to supply the learner with near-immediate feedback on many types of problems. The symbol * is used to indicate "challenge" problems and content that is optional in the sense that it need not be studied in order to maintain continuity.

This book was inspired by the obvious need for a kind of text that uses the best presently known about learning in order to facilitate education regarding mathematically accurate and interesting geometry. Suggests Ruth Wong, "The teacher should have personal involvement with the very mathematical processes through which he will be providing experiences for his students."[1] This is because, as Marguerite Brydegaard says, "Teachers have a tendency to teach in a way similar to that in which they were taught."[2] The author strongly concurs, and herein endeavors to provide such involvement on an adult level. There are also occasional notes pointing out additions and variations helpful to an elementary school teacher.

I owe debts of gratitude for the pioneer efforts of others and for many ideas and phrases that have shaped my thinking. I'd like to express here my appreciation for the writings on geometry that have appeared in recent years, particularly for the admirable booklets of the Nuffield Mathematics Project in England. I wish the authors of these books and articles to know that their works have not gone unnoticed.

Special thanks to Evar D. Nering and Margaret M. Pratt for sharing themselves and some of their ideas, and to Jean Connelly for loyalty, typing, and miscellaneous work necessary in preparing this manuscript.

---

[1] Ruth E. M. Wong, "Geometry Through Inductive Exercises for Elementary Teachers," *The Arithmetic Teacher* (February 1972), p. 91.

[2] Marguerite Brydegaard, "One Point of View," *The Arithmetic Teacher* (February 1972), p. 83.

preface to the student

Your learning is the primary focus of this text. To this end, the text leans on your intuition and your experience with the physical 3-dimensional world in such a way as to make the learning quite painless. In fact, some of you may initially doubt that you are learning much mathematics at all. Recent findings in learning influenced the way in which each chapter is written and organized and influence the text's overall progression from earthly models and multi-leveled explanations to more abstract concepts and concise explanations.

This text leans on your experience and your curiosity in finding things out. By this path, I try to lead you to sound reasoning on content that is mathematically correct. The text places a premium on imaginative thinking—rather than on the accumulation of memorized data—by using observations and generalizations from a familiar context. The lessons learned in doing the exercises you encounter in the body of a chapter are an integral part of the explanation in each section you study. The burden is on you, for the main learning of intuitive geometry comes from structured questions asked the student—and the correlative answers. You will find it invaluable to learning—and getting results on a test—if you try the exercises as soon as you first come to them. Try each exercise before reading on.

The mathematical content is largely determined by that which, intuitively understood, allows an individual to move on to many aspects of worldly existence that rest on a geometrical basis. The point set approach is used throughout. Space is considered as the set of all points; the figures of elementary geometry such as segments, angles, triangles, and circles are subsets of space, as are spheres and solid cubes. If you should founder when encountering a reference to intersection or union of two sets, turn to the appendix; it's for you.

The content assumes only the minimal layman's vocabulary. The first seven chapters could be termed "geometry without measure." With the introduction of Euclidean measure in Chapter 8 and the definition of Euclid's idea of congruence in Chapter 10, the content becomes specifically Euclidean geometry, a metric geometry. Chapters 1–15, comprising the major material of the book, deal with different types of equivalences, culminating in Euclidean equivalence (congruence). You might be interested in the summarizing diagram located at the end of this preface. It is excerpted from the text to show the different shapes equivalent to a given triangle. The progression from topological equivalence to congruence is a part of the material of Chapters 1–15. Then the last two chapters are included to help you tie in the previous material with the realms of coordinate geometry and formal geometry, respectively.

While some of you may have admired the stories of the excited pupils and teachers in such projects as The Madison Project and Nuffield Mathematics Project in England, remember that all the admiration in the world will not alone make us capable of imitating their approach to learning. We have to explore mathematics ourselves from a different viewpoint, and then get into the habit of questioning in place of expounding. Mathematics educators[1] have been asking for just what this book supplies—teacher involvement in learning via experimentation. Those interested in what has succeeded in educating youngsters know that handling, examining, and experimenting with something is quite necessary. It results in learning and questioning, often at a preverbal level. Adults have the same needs, and a chance to "play" with thoughts and objects often precedes an individual's discovery of relationships. For this reason "play chapters," providing units of experience and exposure to new words, are placed strategically about every third chapter. They are meant to

---

[1] Ruth E. M. Wong, "Geometry through Inductive Exercises for Elementary Teachers," *The Arithmetic Teacher* (February 1972), p. 91; Marguerite Brydegaard, "One Point of View," *The Arithmetic Teacher* (February 1972), p. 83.

familiarize you with concepts and vocabulary preceding presentation of related material, not to finalize anything.

There are a few other keys to reading this text, besides knowing how to approach chapters marked "play" chapters. Another special indicator is the symbol *, used to indicate challenge problems or material that is optional in the sense that it need not be studied in order to maintain continuity. You will also notice that there is an answer section at the end of each chapter. I feel that it is important to the learning process to supply you with the near-immediate feedback on many types of problems, so check yourself after applying yourself to an exercise or problem. If an answer does not appear in the answer section, it means that the author wants you to remain challenged.

Have a good time! I hope you find that mathematics can be enjoyable as well as profound.

Gail S. Konkle

A SUMMARY OF THE RELATIONSHIPS BETWEEN DIFFERENT GEOMETRIES STUDIED IN CHAPTERS 1–15

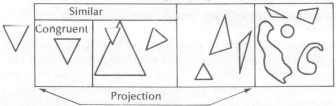

Something topologically equivalent:	any simple closed curve.
Something that is a projection (omitting one special projection):	any triangle.
Something that is similar:	any triangle having sides whose lengths are in the same ratio as the original and having equal respective angles.
Something that is congruent:	one triangle having exactly the same angles and lengths of line segments.

contents

The Architects of Ancient Greece Knew Enough about Perspective to Bend and Tilt the Columns of the Parthenon So That They Would Appear Straight to the Viewer.

Moody Institute of Science

chapter 1 (PLAY)

shapes
&
perceptions

MATERIALS NEEDED

Slide projector and unexposed slide, transparent plastic models of polyhedra

Clay or cheese models of solids and cutting tool

Masking tape and tape measure

Cardboard or wood or plastic patches

   Six congruent square regions
   One large rectangular region
   One large triangular region
   One large circular region

12 inch by 12 inch by 30 inch box

Cylindrical tin and cube circumscribing it

Optional:

Pipe cleaner and colored yarn

Children's blocks

Shoe box

In this chapter you may find yourself involved for the first time with geometry at an intuitive level. This chapter aims to lessen disparities between you and other members of your group by involving you in play ideally woven through your early years. Hopefully, this chapter will help you glimpse the vista of endless possibilities for geometrical exploration. (Your awareness of these possibilities for exploration is one reason for each chapter, indeed, a reason for the whole book.)

FIGURE 1

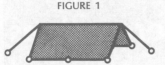

1.1 YOUR PERCEPTION

We can train ourselves to perceive more. In fact, physiologists tell us that the development of the neurological basis for perception is still incomplete in the young child. In addition, psychologists affirm that perception of 2-dimensional diagrams is totally learned—that the connection between an object and its picture is purely a matter of training. We all need this training in order to communicate. Test your comparative absorption of this form of training by completing the picture of the tent in Figure 1.

EXERCISE 1:

We understand in looking at a picture that some lines are hidden from view. Where are all the other lines of the tent in Figure 1 and its loops for tent pegs? Sketch in all remaining lines of the tent and tent loops.

It's very unlikely that you are a "closet case" whose neurological growth was completed without exposure to needed visual experiences. You will see when you check yourself on the following exercise that even a well-developed adult has difficulty in answering some questions requiring visual perception.

EXERCISE 2:

Without using any tool, answer all the questions in Figure 2. When you have committed your answers

FIGURE 2

1.1 YOUR PERCEPTION ⑤ 3

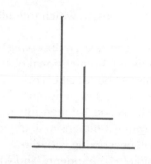

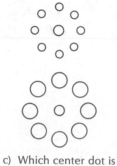

a) Which are equal in length—the lines in the back pair or the lines in the forward pair?

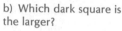

b) Which dark square is the larger?

c) Which center dot is larger?

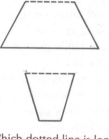

d) Which dotted line is longer?

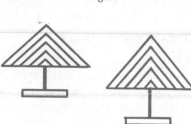

e) In which tree are the height and the width equal?

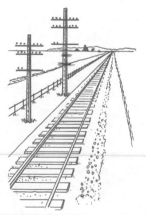

f) Which telephone pole is larger in its environs?

to paper, return to each drawing with a compass, a piece of paper, a ruler, or another tool to check your perception.

You are exceptional if you were right every time, for the context (background) in which you perceive things, and their color and texture influence your sense of size. If you, with perhaps 19 years of helpful conditioning, fell for a few of the tricks on the eye, you can better understand the youngster who, for love or abhorrence of a color or texture, might see the emotionally arousing object as bigger. Our society is filled with standardly recognized illusions; children have to be taught that railroad tracks do not really run together, that slender adults are seldom as tall as they look, that a 10-inch pizza is only half the size of a 14-inch pizza, and that fat children appear fatter when wearing horizontal stripes. Psychology books are filled with optical tricks achieved by switching of the figure and ground. You may not have thought of it before, but doesn't the moon look bigger on the horizon than when overhead? Yet we are essentially the same distance from it. A man named Minnaert wrote a book containing many illusions[1] in which he tells us we can make the moon on the horizon seem the same size as when overhead if, bending over or crawling, we raise our chin so that we have to view the moon with a gaze directed upwards.

[1]Marcel Gilles Josef Minnaert, *The Nature of Light and Colour in the Open Air*, trans. H. M. Kremer-Priest, rev. ed., K. E. Brian Jay (New York: Dover Publications, Inc., 1954), pp. 162–65.

Understanding drawings meant for communication is a matter of training. It seems wiser to engage in frank and unabashed training of ourselves and our young people so that we each may perceive more quickly what someone is trying to communicate and learn to communicate our own perceptions.

In addition to giving practice in perceiving and correctly interpreting drawings, this chapter exposes you to the objects themselves. There is evidence that college students learn geometry best by relating to 3-dimensional objects—as do youngsters. Traditional geometry starts with lines and planes and builds up to 3-dimensional concepts. For years, geometry has been taught from drawings; yet several independent teachers on this continent,[2] and those working in the Nuffield Mathematics Project in Britain have found it more meaningful to the child to first examine objects. This text will make your perception of physical objects a starting point from which to learn about lines, surfaces, measurement, etc. If experiences seem packed into this chapter, it is because your perception is sufficiently developed compared to a child's so that this chapter serves to alert you to specific learnings. The experiences brought into this chapter should spiral through childhood years so that all of us

have highly attuned perceptions on which to build.

In this chapter we will concentrate on observing some geometric properties of 3-dimensional objects. Accompanying this will be specific work on visualizing and decomposing a 3-dimensional object into its 2-dimensional and 1-dimensional components, and on relating 3-dimensional objects to their pictures. We will be working with the physical objects, with simple components, and with drawings.

So now let us begin an interwoven series of questions designed to heighten your awareness of three dimensions and to alert you to your intricate thinking processes. Each of you will find some exercises to stretch your mental and visualizing abilities. Those of you who have studied little geometry may find yourselves in the best position to learn from this chapter and from this entire textbook. This is because many geometry courses breed habitual formula-users, and the formula-user often resists directing his thought to concepts.

1.2 EXPLORATIONS OF SPACE

The emphasis of the exercises in this section is indicated by the capital letter accompanying it. Those exercises marked A, B, and C emphasize the visualization of 3-dimensional objects, including the slicing and counting of faces. In D, you unfold objects to check your ability to visualize a surface. Those exercises lettered E stress refolding and construction of things, and those exercises marked F

[2]Janet M. Black, "Geometry Alive in the Primary Classroom," *The Arithmetic Teacher* (February 1967), pp. 90–93; Brother Alfred Brousseau, "Geometry for the Slow Learner" (Speech given at NCTM Regional Meeting describing work in DeLaSalle High School, Concord, Calif., 1968.)

involve initial explorations in shapes and measurement (comparative volumes in particular), and demand the greatest independence from objects or their representations.

EXERCISE 3A:

Look at a square patch made out of cardboard.

If you had to make a figure enclosing air by taping together duplicates of this patch, could you do it?

If so, how many patches could you use to do this?

EXERCISE 4A:

Can you make a closed figure such as you did in the previous exercise using only duplicates of the cover of this text?

Using a mixture of these and others?

EXERCISE 5A:

Can you make a closed figure out of patches of Figure 3a?

Out of patches of Figure 3b?

Out of patches of Figure 3c?

Answer these questions, making certain that you use patches copied to scale. When your answer is affirmative, proceed to state how many patches it takes to make a closed figure if only one patch lies on the desk top.

For the following set of five exercises, you will be comparing classroom models of a sphere, a cube,

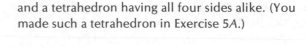

a)   b)   c)

FIGURE 3

and a tetrahedron having all four sides alike. (You made such a tetrahedron in Exercise 5A.)

EXERCISE 6B:

Fill in information in the table below as you examine each of the models listed above. After the name of the model, state how many flat surfaces it has. If it is composed entirely of flat surfaces, fill in how many edges the surface has and how many corners at which the edges meet.

Name of surface	No. of flat surfaces	No. of edges	No. of corners
sphere			
cube			
tetrahedron			

EXERCISE 7B:

Is the number of edges equal to the number of edges on one face multiplied by the number of flat surfaces?

If you were told the number of flat surfaces and the number of corners, could you fill in the number of edges?

EXERCISE 8B:

For each of the three solids, sketch outlines of what flat surfaces you would see if the solid were sliced horizontally as it sits on your desk. (Three ways to check your visualizations: slice clay models, hold the object against a wall and shine a flashlight to make a shadow simulate the cut, or project a thin stream of light on a transparent model from an unexposed slide with a straight scratch on it.)

EXERCISE 9B:

For each of the same solids, sketch the outline you would see if a slice were made vertically through the highest point of the solid as it sits on your desk.

EXERCISE 10B:

Balance a cube on one of its points. Sketch the outline of the surface of each of the following slices: a horizontal slice, a vertical slice through the highest point of the cube, and a vertical slice just left of the high point.

Are the horizontal slices all alike?

Are the vertical slices all alike?

If you have not already done so, you should now be able to complete Exercise 3A. To proceed, join with cellophane tape six identical square patches to form the pattern in Figure 4. This, amongst other patterns, you can fold up into a cube. To aid you in seeing other cube patterns, carry out the following set of instructions (or visualize yourself doing them) with an eye to completing Exercise 11.[3]

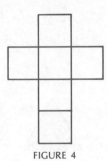

FIGURE 4

Use six squares of cardboard (or plastic) and cellophane tape to make a cube. When you have finished, open it out again without removing all the tape so that the six squares are connected to each other and lie flat on the table.

Try to fold it up again to make a cube. Now use the squares again, and try to arrange them in a different pattern so that they will again make a cube when folded completely.

In how many different ways can you arrange the six squares to make a cube? Copy some of the arrangements onto graph paper to make patterns for cubes.[4]

EXERCISE 11E:

a)   Copy all the patterns you found.

b)   State to a classmate one rule by which you can recognize for certain whether a pattern works or not. (I maintain that you understand the cube better for having to think about patterns you can and cannot fold into a cube.)[5]

While you are contemplating cubes, it might be fun to make a skeleton of a cube (its 12 edges), for

---

Chambers Ltd., 1968; New York: John Wiley & Sons, Inc., 1968), p. 55, by permission of the publishers.

[4] These instructions, using square patches to make a cube, may be used for young people.

[5] It will better enable those of you planning to teach to fill in with your students the learning efforts that precede and accompany the popular game of cutting and folding.

---

[3] Americanization of an assignment suggested in the Nuffield Mathematics Project, *Shape and Size 3* (London: John Murray [Publishers] Ltd., 1968; London: W. & R.

example, using a pipe cleaner, straws, or toothpicks. Several different skeletons hung on colorful yarn make a great educational mobile.

There are classic puzzles in mathematics literature about a spider trying to crawl to a fly, and about a man seeking the shortest route to install wiring in the walls of a room. (See articles by James R. Newman and E. Kasner in the four-volume work called *The World of Mathematics*, [New York: Simon & Schuster, Inc., 1956], p. 2433, and "The Electrical Problem" on page 106 of *Mathematical Puzzles of Sam Loyd*, vol. 2, ed. Martin Gardner [New York: Dover Publications, Inc., 1960].) They all involve finding the shortest distance between points on opposite walls of a room.

EXERCISE 12D:

Here is the classical electrical problem. Along the walls of a meeting hall 30 feet long, 12 feet wide, and 12 feet high, an electrician wishes to run a buzzer wire from front to back walls. Therefore, a loquacious speaker against the back wall can be signalled from the doorway when to stop. The wire must run from 3 feet above the floor on the center of the front wall to 3 feet from the ceiling midway along the back wall. The wire can be strung along walls, floor and ceiling, just so long as it takes the shortest possible route.

a)   What is the shortest route, and how much wire is needed?

(This is not trivial; it's worth your time to experi-

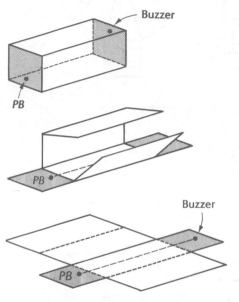

FIGURE 5   This is one way to do it; it is 42 feet this way.

ment with a cardboard box to the scale of the room.) To begin, you can measure various routes with tape measure and examine Figure 5.

b)   Is the shortest path going to cross the same walls if the room is a 12-foot cube?

EXERCISE 13D:

Make a copy of the room and unfold it to check your guess. Can you, by the way you unfold the room, convince yourself of the accuracy of what you now feel in the shortest route?

EXERCISE 14E:

a)   Do you think that all of the patterns in Figure 6 will fold up to enclose air? (Examine each pattern and fold it mentally, answering yes or no.) Pick the one or two most difficult for you to answer, copy each on heavy paper, and try them. Bring your examples of unfolded patterns to class to convince the others whether or not your paper would fold around a solid. (You may have to add tabs and use glue to prove your point.)

b)   Suggest, by coloring each pattern that does fold, a symmetry helpful in the folding. (Use the letter *B* for the blue surface you conceive of as base, then letter the other surfaces of your pattern with such letters as *Y*, *R*, *O*, *G*, and *W*.)

In Exercise 14*E*, we folded patterns we could see and touch. Each pattern is a physical model of the abstract notion of a mathematical surface. A surface has no thickness. Mathematical use of the word

FIGURE 6

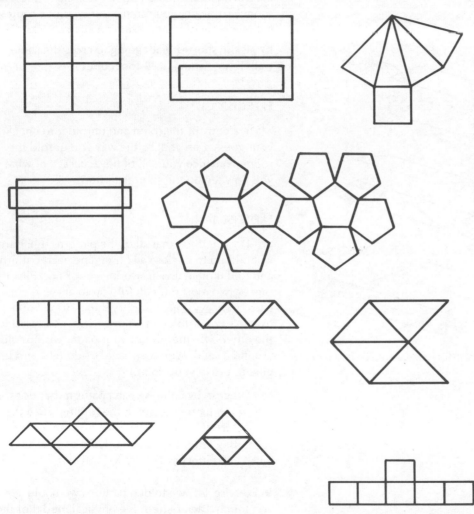

"surface" limits it to a set of points quite different from a piece of cardboard, an orange peel or the top five feet of sea water. We cannot see or touch a surface. A surface actually has no thickness to its walls; no matter how carefully one measures, a surface will neither take up space nor displace water.

EXERCISE 15C:

a)   In a sack made of plastic, cloth, or nylon lingerie, place a big rectangular patch of rigid material. Draw down on the material at a point under the center of the rectangle so that the sack forms a taut closed surface with the rectangle at the top. (See Figure 7).
Notice that you have flat surfaces of rectangular base and four slant sides of the pyramid you have formed. Move the bottom point, preserving the tautness, and you will notice you retain the general inverted pyramid appearance of five flat faces even when the point is far to one side of the rectangular base. All the various pyramids you are forming have a certain set of properties in common: number of flat faces, number of edges, and number of corners.

Copy the table in Figure 8 on a full sheet of paper and complete the first row.

b)   Add a triangular patch to the sack containing a rectangle and push it to the opposite end from the rectangle. Get someone to help hold the rectangle while you pull down on the triangle to make a taut surface. Although crude, this gives you an idea of the surface formed between the rectangle and the triangle. If all faces appear flat, edges will be well-defined, and flat faces, edges, and corners can be counted and inserted in the table in Figure 8. If any surface you investigate in this section is a union of other than flat surfaces, state that it has some curved surface rather than leaving it out of the table.

c)   Repeat Exercises 15a and b using various shapes and combinations of shapes: a rectangular patch, a square patch, a triangular patch, and a circular patch. Extend the chart to list as many as possible.

Discard the plastic or cloth sack as a visual tool whenever you wish. (Throughout this text, feel free to continue an exercise without aids whenever you have familiarized yourself with the aid and no longer need it.)

Bottom

FIGURE 7

Description	Diagram	No. of flat surfaces	No. of edges	No. of corners
Top rectangle Bottom point	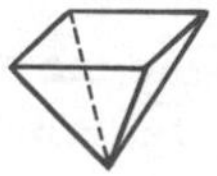	5		
Top rectangle Bottom triangle				

FIGURE 8

EXERCISE 16A:

How good is your imagination? Let's test it! Can you visualize something described verbally? Can you imagine it well enough to count the edges if any? The corners? The surfaces? Test yourself on

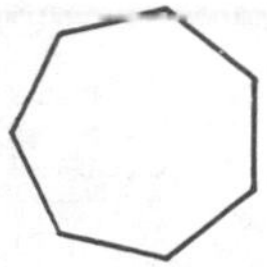

FIGURE 9

the following exercise and enter the information in the table in Figure 10.

a)   Think of a point about 2 inches in the air above the polygon pictured in Figure 9. Now think of the closed surface suggested by joining that point to each of the 7 corners of the polygon. (It is the surface formed by imagining a skin stretched around a rigid wire skeleton.)

Fill in how many flat surfaces this figure has.

If all the surfaces are flat, you should be able to count edges and corners. State the number of edges and corners, if any.

b)   Think of another polygon, just like the one in Figure 9, suspended parallel to your desk but 2 inches above the other. Now imagine the surface formed around the two duplicate polygons directly above each other, connected by vertical lines between the respective corners.

Complete the entries, in the table in Figure 10, for this imagined surface, surface b, and for the previous surface, surface a.

c)   Imagine a surface formed around 2 identical polygons as in part b of this exercise, with the polygon having only 5 sides instead of 7. Fill in the table for this surface, surface c.

d)   Mentally form a surface around a 5-sided polygon and a point somewhere above it, and enter its characteristics in the table. Call this surface surface d.

e)   Mentally form Siamese tetrahedrons, one extending up and the other down from a common base hidden from view. (A tetrahedron is a pyramid of triangular base). Examine your Siamese tetrahedron and fill in the table in Figure 10 (surface e).

Figure	No. of flat surfaces	No. of edges	No. of corners
a			8
b	9	21	
c	7		10
d			
e			

FIGURE 10

Notice that surfaces formed with a circular base or top do not have all flat surfaces. Yet most of the surfaces just constructed do. These are called polyhedra. The polyhedron will be formally defined in Chapter 2, but you are correct in using the word if you are thinking of a closed surface made up of a finite number of flat surfaces.

The polyhedra, because they are similar in nature, display a numerical pattern relating the number of faces, the number of edges, and the number of corners (vertices). Examine the table in Figure 10 for a first glimpse of this pattern before proceeding any further.

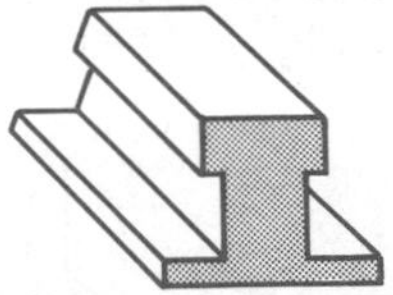

Hopefully, you are justly proud of your increasing ability to visualize 3-dimensional objects. Look now to the fun of reading and creating drawings.

EXERCISE 17B:

Figure 11 shows a block of wood in which a groove has been cut. (Hidden lines are not shown.)

If we were to look at this block from three different positions, which of the three following sketches— 11a, 11b, or 11c—would represent the top view?

The side view?

The front view?

EXERCISE 18B:

a)   Figure 12 shows a piece of railroad track. Make sketches of the top view, the right side view, and the back view of the rail.

b)   Do you think that when a picture is shaded it is easier to comprehend the physical object represented by the picture? Why? Compare Figures 11, 12 and 13 before you attempt to explain your viewpoint.

EXERCISE 19B:

To check your ability to read a drawing and to add three more lines of information to the table in Figure 10, count faces, edges, and corners of each entity in Figure 13. There were five figures already in that table, so let's refer to these three as Figures 13f, 13g and 13h.

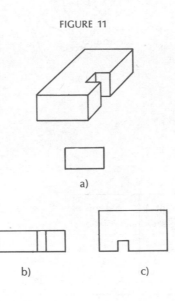

FIGURE 11

a)

b)   c)

FIGURE 12

Try answering all of the following questions without reverting to making a sketch for yourself. Then go back to making a sketch and change your answers as necessary.

EXERCISE 20F:

Could a round can fit inside some cube and touch the surface everywhere?

Could several round cans fit within some cube and touch the cube surface everywhere?

Could several round cans fill up the cube?

EXERCISE 21F:

Consider the same three questions for children's blocks inside a cube. For instance, imagine that someone now hands you a cube 3 inches on a side.

Discuss the possibility of fitting several 2-inch cubes into this cube to fill it.

Discuss fitting 1-inch cubes into this cube to fill it.

EXERCISE 22F:

Consider, instead of a tin can or children's blocks, a box for ski boots.

Could such a box of dimensions $\frac{1}{2}$ foot by 1 foot by 1 foot fit inside any cube and touch the surface everywhere?

Could several be put together to fit this way inside some cube to fill up the cube? If so, state how many boxes it takes to fill up what size cube.

FIGURE 13

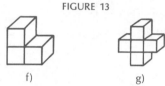

f)   g)

Clustered blocks that are congruent and intersect "nicely"

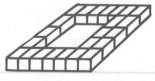

h) Enclosing a Courtyard

EXERCISE 23F:

Repeat Exercise 22F for a ring box having dimensions of 1 inch by 1 inch by 2 inches.

Revert to real objects or careful drawings to check those decisions you question in Exercises 20F through 23F. Now count how many of the answers you had to change. You are truly fantastic if you didn't make a single alteration or only made changes in 1 or 2 places out of more than 13 answers. Rate yourself "less-than-fantastic," "pathetically normal," or "beginner" for erasures needed on three, four and five or more questions, respectively.

The object of this play chapter was to strengthen your intuitive comprehension along avenues that will be developed later in the text.

Although you may have picked up some new vocabulary while playing in this chapter, it is incidental to your happiness right now. Definitions and organized introduction to new concepts will occur in Chapter 2 and all succeeding chapters except those chapters marked "play."

ANSWERS TO EXERCISES

2.   a)   the back one.   b)   same.   c)   same.
     d)   same.   e)   right one.   f)   right one.

3A. 6 or $6 \cdot 2^2$ or $6 \cdot 3^2$; generally $6n^2$ patches used to make cubes.

     Partial List: 10, 14, 16 patches.

4A. Mixtures only; using three pairs of identical rectangular patches.

5A. No, because we specified only one triangle to the plane. (Twelve will do it if two triangles were allowed in the plane.); Yes, using four; No.

6B. Sphere has no flat surfaces;
     for the cube, 6, 12, and 8;
     for the tetrahedron, 4, 6, and 4.

7B. No; No.

8B.

12D. a)   The distance = 41.785 . . . this way, the shortest way.

     b)   The shortest route is of a different nature for the 12-foot room.

13D. Shortest distance between two points is a straight line.

14E. Approximately half will fold.

15C. a)   5, 8, 5 for the rectangular pyramid.

      b)   Only the base is flat.

16A. a)   8 flat surfaces: base plus 7 triangles.
           14 edges
           8 corners

      b)   9 faces: the top and bottom and 7 rectangles
           21 edges: 7 each on top and bottom, 7 vertical
           14 corners: 7 top and 7 bottom.

17B. c, a, and b.

19B. f)   8, 18, 12.   g)   14 faces, 36 edges, 24 corners.

      h)   The fort pictured in Figure 13 has 16 faces, 32 edges, and 16 corners. Its surface is not a polyhedron, and so there is no reason to expect that the number of faces, edges, and vertices fit any number patterns observed on polyhedra. In Chapter 3, we see the difference between this kind of surface (a torus) and the kind of surface in which polyhedra are a subclass. Some of you may have considered the whole top as 1 face; if so, you logically got the answer 10, 24, 16. In Chapter 3, the word "face" is defined so that you can understand why this fort is said to have 16 faces.

20F. No; No; No.

A Cylinder Within a Cube

21F. Yes to the 3 questions for blocks of the right size; Cannot do it; Can do it with 27 cubes.

22F. No; Yes, 2 of them to fill up a 1-foot cube.

23F. No; Yes; Yes, 4 boxes will fill up a 2-inch cube.

Black, Janet. "Geometry Alive in the Primary Classroom." *The Arithmetic Teacher.* (February 1967), pp. 90–93.

Brousseau, Brother Alfred. "Geometry for the Slow Learner." Speech given at NTCM Regional Meeting, December 1968, describing work in DeLaSalle High School, Concord, California.

Loyd, Samuel. *Mathematical Puzzles of Sam Loyd.* Vol. 2. Edited by Martin Gardner. New York: Dover Publications, Inc., 1960.

Minnaert, Marcel Gilles Josef *The Nature of Light and Colour in the Open Air.* Translated by H. M. Kremer-Priest. Revised by K. E. Brain Jay. New York: Dover Publications, Inc., 1954, pp. 162–3.

Newman, James R., ed. *The World of Mathematics.* Vol. 4. New York: Simon and Schuster, Inc., 1956, pp. 2433–34.

Newbury, N. F. "Quantitative Aspects of Science at the Primary Stage." *The Arithmetic Teacher* (December 1967), pp. 641–4.

Ptak, Diane. M. *Geometric Excursion.* Oakland County Mathematics Project. Oakland, Mich.: 1970, pp. 24–32, 48–59, 62–86.

Smith, L. B. "Geometry, Yes–But How? "*The Arithmetic Teacher* (February 1967).

Walter, Marion. "An Example of Informal Geometry: Mirror Cards." *The Arithmetic Teacher* (October 1966).

Wardrop, R. F. "A Look at Nets of Cubes." *The Arithmetic Teacher* (February 1970), pp. 127–8.

Wenninger, Magnus J. "Polyhedron Models for the Classroom." *National Council of Teachers of Mathematics.* Washington, D.C., 1966.

Abbott, Janet S. *Learn to Fold—Fold to Learn.* Pasadena, Calif.: Franklin Publications, Inc., 1968.

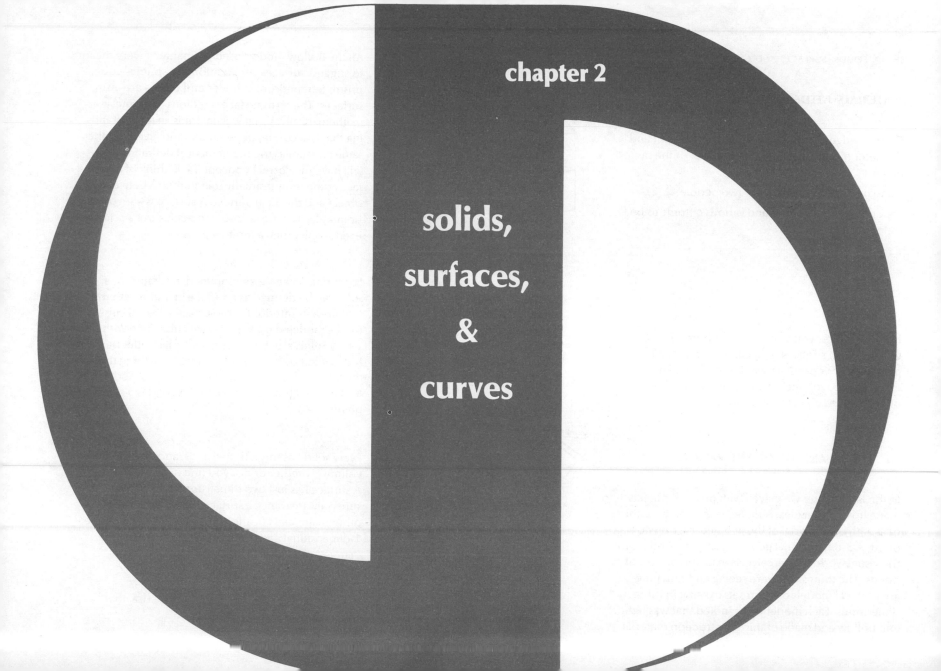

chapter 2

solids,
surfaces,
&
curves

MATERIALS NEEDED

Cube of edge *d*, sphere and cylinder of height and diameter *d*, solid ball and cubical block of the same height

Hand mirrors for each one or two people

Paper polygonal regions and various cutouts to be tested for symmetry

In this chapter, you will learn to differentiate between solids, surfaces, and curves. An initial explanation of vocabulary will be reinforced as we explore size (volume) of space objects and symmetry of both space objects and curves.

2.1 THE MEANINGS OF THE WORDS

In the last chapter we played with physical objects made up of molecules. A molecule can be thought of as a physical model of the geometrical object, a point. A 3-dimensional geometry, corresponding to the visual world we perceive, is made up of a set of points. The things you were seeing and touching are physical models of point sets existing in three dimensions. Each model you handled that was light and hollow and made of thin construction material such as cardboard was intended to suggest a surface.

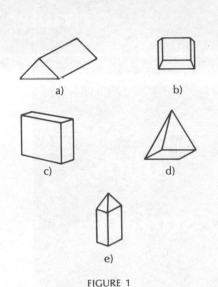

FIGURE 1

FIGURE 2

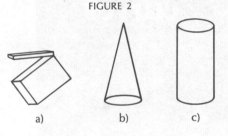

d) Paper Fan

All the hollow models used in Chapter 1 were meant to suggest surfaces; in addition the sphere, cube, prism, tetrahedron, cylinder and others are also surfaces. The term surface is tedious to define mathematically. What we aim for is an understanding that is accurate, dependable, and gives you the same concept as the mathematical definition. You will have the correct concept if you think of a *surface* as the infinitesimally thin interface between an object and the air around it. A surface is a set of points like a single layer of molecules, yet a point is even smaller than a molecule.

For surfaces like those examined in Chapter 1, a *solid* can be defined as a surface in union with the set of points interior to the surface. (A solid cube can be modeled by a cardboard cube filled with sand.) Solids are those mathematical entities perfectly modeled by all physical objects. Anything that has thickness can model a solid: a piece of wood, a ball, a wedge of cheese, a jackknife. A solid is a set of points.

Every solid has three dimensions, and thus has volume, whether or not you find it easy to measure. A surface has just two dimensions, and we can measure only area since a surface has no depth. The other major classifications of geometric objects are 1-dimensional and 0-dimensional: they are called "curve" and "point," respectively.

There are some surfaces, called polyhedra, that

geometers mention frequently. The surfaces in Figure 1 are polyhedra.[1]

The surfaces in Figure 2 are not polyhedra.

EXERCISE 1:

The surface of which of the solids in Figure 3 is a polyhedron?

A *polyhedron* can be thought of as a "closed" surface, having all flat sides (faces)—or as a boundary of a solid having all flat sides. A polyhedron is a set of points. In Figure 4 are some terms to help you discuss the following questions.

EXERCISE 2:

If a figure is a polyhedron, what must be true of its faces?

Of its edges?

EXERCISE 3:

Could a polyhedron weigh anything?

EXERCISE 4:

Could a face of a polyhedron be modeled by a piece of an egg shell?

---

[1]Some authors allow a polyhedron to be open, in which case these would be called closed polyhedra.

FIGURE 3

a) Orange Crate

b) Solid Ball

c) Modern Stool

d) Kitchen Table

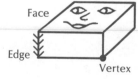

FIGURE 4

EXERCISE 5:

What is the least number of corners (vertices) a polyhedron can have?

EXERCISE 6:

What is the greatest number of faces a polyhedron can have?

The faces of a polyhedron are each surfaces—and flat surfaces at that. A flat surface is called a *plane* or a piece of a plane. A face of a polyhedron and the surface of your desk top are each pieces of a plane. Unless differentiation is needed, both whole planes and pieces of planes are referred to as plane surfaces. Using the word "plane," we can state a better definition of a polyhedron. A *polyhedron*, then, is a "closed" surface having all plane faces. Other surfaces than plane ones are not found on polyhedra, but we see them all the time. Some curved surfaces are those of an egg, a football, a globe, and a cone.

Each face of a polyhedron is polygonal; that means that the outline of each face is a polygon. As an example, look at Figure 4 to affirm that the polygon outlined by each face is a rectangle. A polygon is a curve, as are line segments, rays and angles. A few names of special curves are important to know at this point.

Each edge of a polyhedron's face is called a *line segment*. A line segment has two endpoints. A *line*, however, is similarly straight and is described by a mathematician as extending infinitely in both directions; by this he means that the line is not of finite length and so cannot be measured at all. (He does *not* mean that the length is a very big number.) A ray is a "half line," written $\overrightarrow{pq}$.

A line cannot be drawn on the blackboard, but it can be suggested by placing arrows at each end. Hence, you could mark

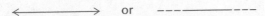   or   — — — — — — — — —

on the blackboard in answer to the question, "What is the set of all points which lie in the plane of the teacher's desktop and in the plane of the blackboard?" All point sets of 1 dimension are classified mathematically as curves. The line and the line segment are particular curves. Other 1-dimensional sets of points are usually referred to merely as curves, but special vocabulary is retained for such curves as the angle, the circle, the rectangle, and the triangle.

We will now have to use the word "intersection" to communicate. If you feel insufficiently familiar with sets of points and intersection of sets of points, turn to the appendix.

EXERCISE 7:

In Chapter 1, you sliced surfaces and solids. A slice of a surface is the intersection of a plane and a surface. A slice of a solid, often called a "plane section" or "cross section," is similarly the intersection of a plane and a solid. Let us confine our attention to those slices that touch at more than one point, and consider the resulting kinds of slices.

a)   If two different planes intersect, then their intersection is a line.

Draw the set (or sets) of points we get when intersecting a plane and a sphere.

When intersecting a plane and a rectangular prism (box-shaped surface).

Circle the type(s) of point sets that result from the intersection of any plane and a surface different from it:

solid   surface   curve

b)   Draw the set (or sets) of points representing the intersection of a solid ball and a plane.

Of a plane and a wedge of cheese.

Circle the type(s) of point sets that result from the intersection of any plane and a solid:

solid   surface   curve

You can rapidly become accustomed to this point set phraseology by browsing through grade school mathematics books and reading certain articles on teaching geometry. The latter are most numerous in a journal called *The Arithmetic Teacher*. One such article is "Geometry Yes—But How?" by L. B. Smith, found in *The Arithmetic Teacher*, February 1967. One device he uses to clarify vocabulary for sets of points is to letter the faces of a physical object (such as a polyhedron model) and to ask the students to make a mark representing a point in surface *S*, then several marks in surface *R*, then several representing points of intersection of surfaces *S* and *R*. Can you do the same on any of the solids or closed surfaces discussed in Chapter 1? At another time, Mr. Smith asked the students to put chalk marks on the chalkboard indicating points of intersection of the chalkboard and the plane containing the top of a cube standing on a table; the set of all points in the intersection was then decided upon as a line segment running the full length of the chalkboard.

The same article also helps to get across the numerousness of points in geometry. An infinite number of points can be clustered on a pencil mark. This is due to the nature of a point. A point can be thought of as the limit in preciseness to describe a location in space. Note the successive preciseness:

I am on Canal Street.
I am in my house.
I am in my bedroom.
I am in the left rear corner of my bedroom.
etc.

As your position becomes more and more precisely expressed, the geometric center of you approaches a point in space. A point is an idealized location in space. It can be thought of as a position on a line located in space. A point really has no dimension at all—no length, no width, no depth. Since it takes up no space, it cannot be the same as a chalk mark or a pencil dot; yet these are rightly employed to represent a point or to indicate the position of a point.

You have just been introduced to some of the most important vocabulary. It is crucial, in to order to communicate, to know the difference, for instance, between a solid whose model has weight and thickness and a surface where questions about either would be meaningless. ◑

◑ Note To Prospective Teacher:
Differences between solids and surfaces can be subtly communicated by a teacher's cues. You can suggest a surface by sweeping the palm of your hand parallel to the set of points referred to.

This   FIGURE 5   Not This

Another way to communicate surface is to use a shadow because it so obviously lacks depth and yet can be seen. If an object is properly oriented and is not too complicated, a useful shadow will be thrown when a flashlight beam is directed to one side of this object. Under these circumstances, a plane surface of an object can be perfectly copied on the wall of a darkened classroom.

FIGURE 6
3 Shadows of a Toothpick Box

Side Face   Combination showing more than a single face   Top Face

End of Tin Can

Note to Prospective Teacher: (Suggestion)

Sun show or Halloween shadow quiz: (after the mention of flat surfaces and shadows.)

Have a small team prepare materials for a sun (flashlight) show of the shadows cast on the ground (wall) of two or three polyhedra. Have the team build one polyhedron from a carton. Conduct a short quiz on the number of edges each shadow will have when the box is held in several positions, then go into the sun to check answers. The shadows will amount to plane sections like those of Exercise 7b.

EXERCISE 8:

Some surfaces are simple to describe verbally. A sphere can be described as the set of all points that are a certain distance from a point.

Now discover and write down several simple descriptions of a cylinder. (See Figure 7) Perhaps the following excerpts from Franklin Publications' supplementary workbook[2] will give you an idea (Figures 8 and 9); but note that the workbook demonstrates pieces of cylinders rather than cylinders, for that is an easier way to introduce the concept.

EXERCISE 9:

a) Draw two pictures of surfaces formed from a rectangle and a pencil tracing the rectangle, each on the order of the cylinders in Franklin Publications' workbook. (See Figure 8)

b) Draw two pictures of surfaces formed from a rectangle and a point somewhere above it. (The surface is that one swept out by the pencil when kept in contact with the point as it traces the rectangle.)

c) Draw two pictures of surfaces formed from a circle and a point somewhere above it.

[2]Susan Roper, *Paper and Pencil Geometry* (Pasadena, Calif.: Franklin Publications, Inc., 1966), pp. 93–94, by permission of the publisher.

FIGURE 7

Here are some named surfaces alongside a corresponding physical object frequently used to demonstrate that surface.

Rectangular Prism   Cube

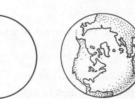

Cylinder   Sphere

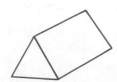

Triangular Prism   Cone

FIGURE 8

2.2 FEELING FOR SIZE ⊕ 21

Place your pencil point on any point of the circle below. Hold your pencil so it forms a right angle with the page. Now slowly move your pencil around the circle. The points in space through which your pencil moves mark the space figure we call a right cylinder.

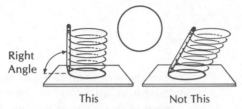

Right
Angle

This   Not This

Here are two pictures of a right cylinder.

FIGURE 9

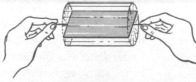

Another way to picture a right cylinder is to rotate a rectangle. Cut out a rectangular region and paste a toothpick on each end where marked.

Now spin the rectangle by twisting the toothpicks between your thumb and index fingers. Do you see the cylinder?

2.2 FEELING FOR SIZE

Sense of size is frequently affected by such things as color, texture, and weight. Mixed subconsciously with the child's idea of size are many irrelevant properties of an object. It is possible that early experiences in considering the volume of an object or the volume needed to fill an object will help isolate the question of size from other interesting properties of an object. Perhaps a fresh look at volume will also relieve you of some misconceptions.

EXERCISE 10:

Form a mental image of two solids on a table side by side—a sphere 2 inches across and a cube sitting flat and measuring the same 2 inches across the face toward you.

a)   If each had only a negligible wall thickness, which of the two do you think would contain the largest quantity of water?

Greater by just a fraction?

Twice the quantity?

Guess as close as you can by mental visualization.

b)   If you cannot be sure of your answer to part a, test it with models of such surfaces of like breadth. (Find a hollow ball and fashion an accompanying box and liner.) If you make an opening in each, you will be able to compare their volumes by pouring sand or liquid from one into the other.

c)   Now that you know for sure which has more

volume, can you think of any reasoning by which the larger was obviously so? State your reasoning on this.

d)   If you had a solid ball and a block of the same breadth, you could use a method of comparison suggested by Archimides in his "Eureka discovery." Describe exactly what you would do to demonstrate this method to others in the class; be prepared to demonstrate that the solid cube has a volume about twice that of the sphere of same breadth. (For practical purposes, the Archimidean demonstration that the cube holds a greater volume will work when the sphere has diameter up to 1.4 times the face of the cube.)

EXERCISE 11:

Visualize a tin can that measures 2 inches across and is 2 inches high. Compare the volume of this tin can with that of the 2-inch cube and the 2-inch sphere compared in the previous exercise.

a)   State your expectations as to what you will find when you experiment by filling or by using the Archimidean method. Explain your reasoning.

b)   If you are not sure of your reasoning, form a mental picture of the tin can placed in front of the cube and the sphere respectively. Carry your reasoning through by using real objects if necessary.

c)   By now you may be so sure of yourself that you can rank the three objects on a logical basis. If so, try your explanation on a fellow student using words and a blackboard or transparencies.

A Hexahedron
FIGURE 10

d)   If you don't feel completely successful with part c, then write out the steps and results for either a sand or water demonstration or an Archimidean demonstration; be prepared to do your demonstration before others if called on in class.

2.3   SYMMETRY

In Exercise 4 of Chapter 1 you mentally joined two tetrahedrons on a common base to make a closed surface of six triangular patches. If both tetrahedrons have identical equilateral triangles bounding all four faces, then we would say that the tetrahedrons are placed symmetrically about the common base, triangle *ABC*. We would also say that the hexahedron created in Figure 10 is symmetrical about the plane of triangle *ABC*. _Symmetry about a plane_ can be thought of as a mirror image placement of points; if the edge of a hand mirror is held against a model hexahedron along *AC* in the plane *ABC*, then peering in the glass gives the impression of a continuation of the bottom half of the hexahedron that is hidden. This mirror effect can only be seen when the points above the plane of the mirror are really symmetrical to those below. If the two tetrahedra are lopsided, one leftward and the other toward the right, then a mirror at the common base will not present one as a reflection of the other. Try this experiment with paired tetrahedra that lean opposite ways and with a pair that lean in the same direction; then you will better understand why the latter makes a symmetrical hexahedron while the former is non-symmetrical.

Symmetrical objects	Non-symmetrical objects
globe, through equator or any great circle	drinking glass, through all but centered vertical planes
drinking glass, through any "centered" vertical plane	typewriter, through any plane
book, through many planes	teapot, through all but one vertical plane

Although the concept of symmetry can be more exactly defined following a discussion of length in Chapter 9, a feeling for symmetry will be conveyed here, by example and verbosity.

Symmetry can be intuited well through examination of earthly objects. Solids and surfaces have the possibility of being *symmetrical about a plane*. Try holding something like a pop can against a mirror, moving your head until you see the original and the image sitting side by side. The "Siamese can" you perceive is said to be symmetrical. It is also proper to say that the original can appears symmetrical to the reflection. As you pull the can further and further away from the mirror, you continue to see two cans symmetrically placed because of the position of the mirror seemingly halfway between the two cans; we say that the mirror acts as a *plane of symmetry* for the two cans. We see objects symmetrically placed in homes. Matching bedside tables are usually placed symmetrical to a

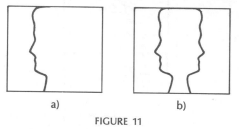

a)   b)

FIGURE 11

vertical plane running up the center of a double bed or running halfway between a pair of twin beds; the pair of end tables is said to be *symmetrical* because, point for point, a mirror image of one is created on the other side of the plane of symmetry.

It is also possible for solids and surfaces to be symmetrical about a line or symmetrical about a point. These are two separate considerations. Every drinking glass I've ever seen has a line of symmetry down the center, but seldom has a point of symmetry. Globes and footballs each have a point of symmetry at the center. You may already intuit the meanings of the two phrases, "symmetry about a line," and "symmetry about a point." To furnish a double check and to make these ideas concrete, we shall discuss one type of symmetry at a time and start with sketchable sets of points in the plane.

One way to check for *symmetry about a line* is to place the edge of a mirror on the line or to measure perpendicularly across the line to check for the presence of *twin points*. The mirror usage popular in modern grade-school workbooks is a great aid in learning about line symmetry.[3] Cards called "mirror cards" can be matched only when the use of a mirror on one would yield the other. For example, can one, by using a mirror on the card shown in Figure 11a, see the pattern shown on the card in Figure 11b?

[3] Marion Walter, "An Example of Informal Geometry: Mirror Cards," *The Arithmetic Teacher* (October 1966), pp. 448–62.

Yes, you can get Figure 11*b* from Figure 11*a* by placing the edge of the mirror on a vertical line.

EXERCISE 12:

Which mirror cards can you obtain by using a mirror on Figure 12*a*?

EXERCISE 13:

Put a line where the edge of the mirror must be placed in Figure 13*a* in order to see Figure 13*b*. You can see quite a few different pictures from holding mirrors on Figure 13*a*; if you desire more practice, continue indicating the resulting view for each line you use.

If a figure yields itself when a mirror is used, then the figure has symmetry about the line the mirror follows. The line along which you place the mirror is called the *axis of symmetry* (or line of symmetry). If you were thorough, you noted that Figures 12*b*,e and 13*b* have lines of symmetry.

Folding is another easy way to check for line symmetry in a plane figure. You have found an axis of symmetry for a piece of paper if you succeed in folding the sheet so that its edges coincide. Cutouts of things such as triangles, parallelograms and paper dolls, should also be investigated to determine whether there is an axis (or many axes) of symmetry for each. An equilateral triangle, for example, has three axes of symmetry; there are three different folds by which opposite points coincide.

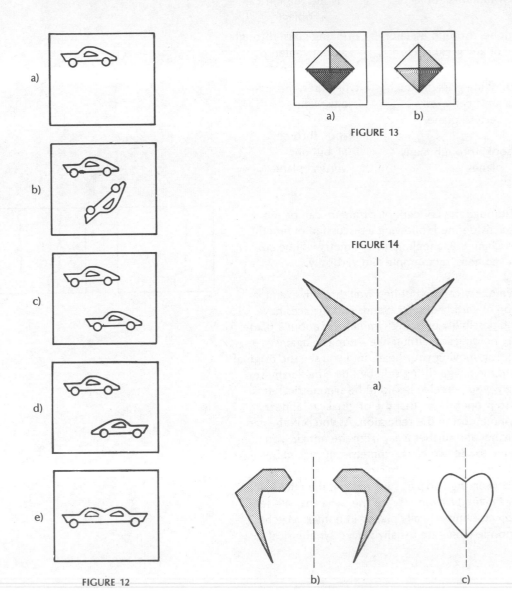

FIGURE 12

FIGURE 13

FIGURE 14

EXERCISE 14:

The idea of folding to check for line symmetry suggests that distance across the line is the issue.

a)   Cover up the left half of each picture in Figure 14 with a blank sheet of paper and test your ability to recreate the left surface using the vertical line, the edge of a blank sheet of paper, and a ruler to help you.

b)   Draw a curve at random. Now draw a horizontal line and then draw that which is symmetrical to the curve about that horizontal line.

Since *symmetry about a line* depends only on the presence of twin points that match across a line, there will be 3-dimensional surfaces and solids with *line symmetry*. Bottles, tin cans, telephone poles and boxes are examples of common shapes having line symmetry.

Symmetry about a point can exist independently of line symmetry. Yet a figure such as a circle has both. A circle is a glutton for symmetry. It has an infinite number of axes of symmetry and a point of symmetry where all the diameters intersect. A figure having a point of symmetry need not be quite as balanced-looking as a circle. Notice that the two surfaces in Figure 15a are alike in some sense; examination of these surfaces with a straightedge will confirm that corresponding points are opposite each other along a straightedge passing through the same point—the *point of symmetry* (which lies on the halfway line sketched in Figure 15a).

FIGURE 15

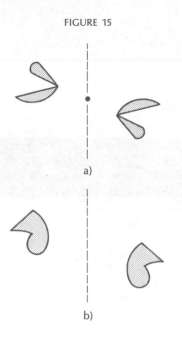

a)

b)

FIGURE 16

EXERCISE 15:

a)   For each picture in Figure 15, cover up the left half with a blank sheet of paper and recreate the left surface when the edge of the paper is along the dotted line. Use a ruler to help you, and denote the point of symmetry on the paper's edge. (Figure 15b should be the trickier, for you have to find the point of symmetry before you can follow these directions.)

b)   Draw any curve, and place a point somewhere on your paper. Now draw a curve symmetrical to the original about the point you chose.

EXERCISE 16:

Thinking in 3 dimensions, consider what else could have a point of symmetry.

Can something solid be symmetrical about a point? Back up an argument with specific examples.

Can some surface other than that capable of containment in a plane be symmetrical about a point? Give two examples backing your view point.

Symmetry is a necessary prerequisite to counting the faces and edges of a solid or of a closed surface when looking from afar or looking at a picture. Can you tell by looking at Figure 16 whether there are 4 missing faces or just 2 or just one? Probably not. It is almost as difficult to tell if someone holds up a solid or projects its shadow to the front of your classroom. But you could reply regarding the appearance of the whole object if you were told that the back of the solid is symmetrical to the part you are seeing. This solid, then, is said to be symmetrical about a plane through the top point and the outside of the visible lower vertices. (You can easily indicate the plane of symmetry for a solid you hold by slicing your free hand through the air.)

EXERCISE 17:

Prepare polyhedra that are symmetrical about some plane. Each member of the class should prepare his own model of a polyhedron. In class, each person files past the front of the room, shows only one of the symmetrical halves of his model and indicates the plane about which the model is symmetric; meanwhile seated members of the class work to fill in on paper all of the pieces of the table in Figure 17. Afterwards compare answers with those given by the student who showed the polyhedron.

⤶ Note to Prospective Teacher:
Exercise 20 is excellent for children if pointing and various handmotions are allowed.

FIGURE 17

Name of person holding model	No. of faces	No. of edges	No. of vertices

2.4 EXERCISES FOR CHAPTER 2

18. a) Draw a sketch of two lines in the plane, one line acting as an axis of symmetry for the other.

   b) Draw a sketch of two lines in the plane, both lines having the same axis of symmetry.

19. Would a round layer cake with smooth outside surfaces be symmetrical with respect to one or more planes? Describe.

20. Describe verbally the following sets of points in space. ⊕
   a) The points not on a cube.
   b) A cube and its interior.
   c) All points not on a cube nor in its interior (complement of a cube and its interior).
   d) Complement of a sphere.

21. Draw 2 points and the set of all points on a line passing through these 2 points. This begins a pattern of pictures. For the next picture, draw 3 points and the set of all points on lines passing through pairs of points when no 3 points are on the same line; repeat for 4 points, no 3 of which lie on the same line; continue this for 5 points, for 6.

By counting the number of lines in each picture, you may be able to guess the number of lines passing through n such points. You'll be getting there if you can make guesses as to the number of lines through 10 points and 100 points, even if you have trouble writing in the number which accompanies n.

FIGURE 18

No. of points	2	3	4	5	6	10	100	n
No. of lines								

1. 3c and 3d are polyhedra, the only polyhedra.

2. Flat faces and straight edges.

3. No; because a polyhedron does not even have the thickness of molecules.

4. No; because each face must be flat everywhere.

5. Four vertices.

6. Any number of faces, providing there are at least four.

7. a) Plane and sphere always intersect in a circle. A plane and a rectangular prism intersect in a triangle, a quadrilateral, a pentagon, or a hexagon. A plane intersects any surface in a curve.

   b) A plane intersects a solid sphere in a circular surface; a plane intersects a cheese wedge in a triangular patch or in a rectangular patch or in one of many other surfaces. The intersection is always a surface.

9. a)   b)

11. a) Volume of sphere < volume of cylinder < volume of cube.

    c) Cylinder touches the bounding cube in a full 1-inch radius circle on top and bottom, whereas sphere has the same radius at center while narrowing to a point at top and bottom surfaces of the cube.

12. Figures *b* and *e* are mirror images of Figure 12*a*.

13.

16. Symmetry about a center point exists in the solid ball—and in the surface of ball (globe) too. One can intuitively affirm this symmetry by noting that any plane passing through center is a plane of symmetry. (Our equator is not the only example; each longitude circle, from the Greenwich meridian on around, determines a plane of symmetry passing through the center of the earth.) And, there is symmetry of points on each of these planes through the center. Hence it can be argued that every point on a globe has its counterpart in a point on the opposite side of a line through the center. Other examples of center symmetry include a cube and a shoe box and a book (sans lettering), two surfaces and a solid.

18. a)   b) ═══

19. Symmetrical about each vertical plane through center, and symmetrical about the plane between layers if there is no frosting.

20. a) The points inside the cube union the points outside the cube.

    b) Solid cube, *c*.

21. 2 + 1 lines through 3 points.
    3 + (2 + 1) lines through 4 points.

You continue the pattern so that you can fill out the table in Figure 18.

BIBLIOGRAPHY

Abbott, E. A. *Flatland: A Romance in Many Dimensions*. 5th rev. ed. New York: Barnes and Noble Books, 1963.

Abbott, Janet S. *Mirror Magic*. Pasadena, Calif.: Franklin Publications, Inc., 1968.

Copeland, Richard W. *Mathematics and the Elementary Teacher*. Philadelphia: W. B. Saunders Co., 1966, pp. 264–7.

Dotson, W. G. Jr. "On the Shape of Plane Curves." *The Mathematics Teacher* (February 1969), pp. 91–94.

Garstens, Helen L., and Jackson, Stanley B. *Mathematics for Elementary School Teachers*. New York: The Macmillan Co., 1967, pp. 119–24, 131–5.

Johnson, Donovan A. "Paper Folding for the Mathematics Class." *National Council of Teachers of Mathematics*. Washington, D.C., 1957.

McFarland, Dora, and Lewis, Eunice M. *Introduction to Modern Mathematics*. Boston: D.C. Heath Co., 1966, pp. 70–83.

Nuffield Mathematics Project. *Shape and Size 3*. New York: John Wiley & Sons, Inc., 1968, pp. 23–25, 44.

Ptak, Diane M. *Geometric Excursion*. Oakland County Mathematics Project. Oakland, Mich., 1970, pp. 7–19.

Richards, Pauline L. "Teaching Geometry Through Creative Movement, Tinkertoy Geometry." *The Arithmetic Teacher* (October 1967), pp. 468–9.

Roper, Susan. *Paper and Pencil Geometry*. Pasadena, Calif.: Franklin Publications, Inc., 1966, pp. 93–94.

Schaaf, William L. *Basic Concepts of Elementary Mathematics*. New York: John Wiley & Sons, Inc., 1966, pp. 234–43, 80.

Smith, L . B. "Geometry Yes—But How? *The Arithmetic Teacher* (February 1967).

Walter, Marion, "An Example of Informal Geometry: Mirror Cards." *The Arithmetic Teacher* (October 1966).

Wenninger, Magnus J. "Polyhedron Models for the Classroom." *National Council of Teachers of Mathematics*. Washington, D.C., 1966.

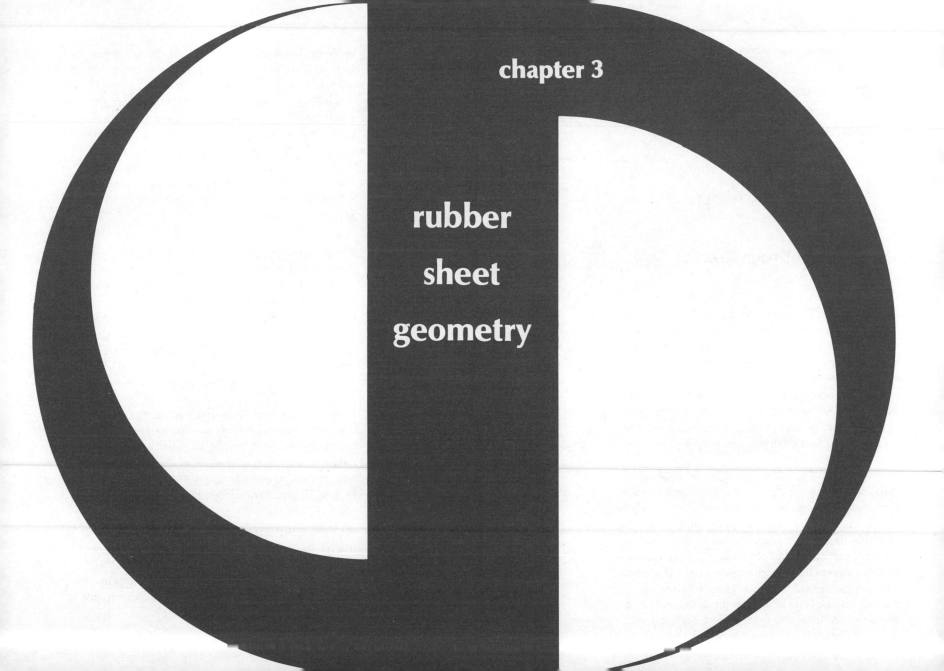

chapter 3

rubber
sheet
geometry

MATERIALS NEEDED

Individual rubber bands, balloons and pieces of string

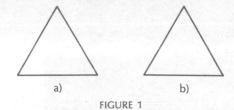

a)   b)

FIGURE 1

3.1 TOPOLOGICAL EQUIVALENCE

The following chapter is an introduction to topology, a "loose" geometry of figures. These figures can be transformed into each other by stretching and bending, much the same as wet clay is molded without being torn. In this geometry, a curve drawn on a balloon, for example, would be equivalent to any curve obtained by stretching the balloon. After defining topological equivalence more fully, we will discuss some topological properties shared by equivalent surfaces, curves, and solids.

In a high school geometry book, one usually finds many occurences of the word "congruent" and its symbol, ≅. The symbol has two parts, = and ~, the former standing for equal and the latter for similar, referring to size and shape, respectively. Congruency is a particular kind of geometrical equivalence. (It is the kind of equivalence associated with Euclid.) Two things are said to be congruent if they coincide when one is picked up and placed on top of the other. Perhaps you remember a high school geometry teacher stressing the properties two tri-

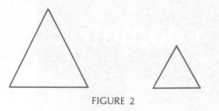

FIGURE 2

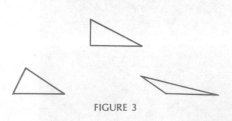

FIGURE 3

angles must have in common in order to be congruent; he might have told you that triangle $ABC \cong$ triangle CDE if line segments $\overline{AB}$ and $\overline{CD}$ are of equal length, if line segments $\overline{BC}$ and $\overline{DE}$ are of equal length and if the angle between the compared line segments has the same measure in the respective triangles. When you contemplate the sophistication of measuring involved in only that check on congruency, you realize that congruency is a rather advanced concept relating to exact sizes and shapes of things.

Can you imagine a geometry book without the word "congruent" in it? Try visualizing the set of figures still equivalent to a certain triangle after we strip away some of the conditions of congruence. Consider a triangle 2 centimeters on a side. (Figure 1a). The congruent figure to its right has the same size and shape. You will note that the following figures appear successively less like the original as we relax requisites of geometrical equivalence.

Now relax the condition that equivalence depends upon the triangles having the same size, and keep the conditions that they be triangles and that they have the same shape. If we omit the condition of size, we still have a triangle with all sides of equal length. (Figure 2 shows two.)

Now disregard shape enough to allow the second triangle to be just any triangle. (Figure 3). The word "triangle" specifies its shape, though not as definitely as when we demanded that the sides be equal. By saying "triangle," we are still specifying that the figure have three sides, each of which is a line segment.

Now remove the condition that equivalence necessitates that the sides be line segments, and allow them to be any arcs, and you can get figures like these in Figure 4. These figures can, by a stretch of the imagination, be considered as having three sides that are not all line segments.

Relax the condition that the curves have three sides and you get such curves as in Figure 5. Figure 5 shows the epitome of freedom, figures "*topologically equivalent*" to the original 2-centimeter triangle.

What do the nine curves in Figures 4 and 5 have in common with the original triangle? Certainly they are more like that triangle than those in Figure 6. A rubber band model of that triangle can be stretched and pulled into the form of any of the examples shown in Figures 1 through 5, but not into any in Figure 6 without crossing the rubber band. (If the original triangle were drawn on an infinitely stretchable sheet of rubber, the same would be true: there would be a certain stretching equivalence between the Figures 1 through 5, not shared with those in Figure 6.) Try penning a triangle like that in Figure 1 on a piece of burst balloon or on some other elastic sheet, and stretch the material. You will find that with good use of your fingers—and maybe those of a friend—you can stretch your rubber sheet figure into any size triangle (Figure 2), into any triangle (Figure 3) and into any of the

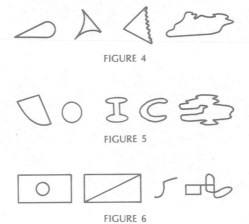

FIGURE 4

FIGURE 5

FIGURE 6

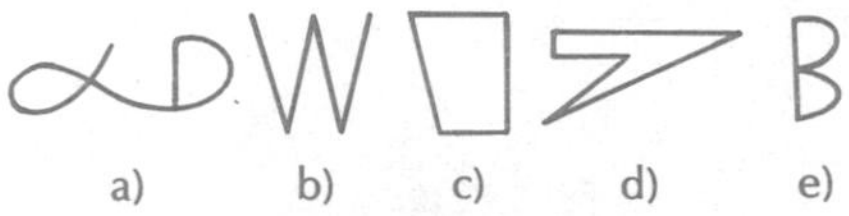

a)   b)   c)   d)   e)

FIGURE 7

⑩ Note to Prospective Teacher:

The child of today is introduced to topological properties very early in his school career. That the study of topology involves fundamental geometrical concepts is one reason for this; one needs fewer background concepts before meaningful exploration can begin. Furthermore, work with figures depends less upon visual and manual experience and sophistication because exact shape and size do not enter in.

curves in Figures 4 and 5, all of which are topologically equivalent.

EXERCISE 1:

Put a check mark beneath each of the curves in Figure 7 you deem topologically equivalent to those in Figures 1 through 5. (Use the primitive idea of elastic motions; a more complete definition follows.)

This kind of geometrical equivalence has fewer conditions than does the equivalence called Euclidean congruence. (Obviously! We just wiped away many of the conditions.) It is called topological equivalence.

Because the properties shared by topologically equivalent figures are those maintained through an elastic motion, *topology*, the study of these properties, is nicknamed rubber sheet geometry. Topology has fewer rules to concern the player. Those few properties considered to be worth maintaining in topology are also things Euclid deemed important; they are basic to all the geometries we will study. The topological properties of a figure are the most fundamental of its geometric properties—ones on which we build. ⑩

This text will follow a generally logical development, examining topological equivalence, building from this to increasingly more specialized geometries—just the reverse of the stripping away process performed on the 2-centimeter triangle.

3.2 TOPOLOGICAL PROPERTIES

Topology is the study of topological properties of
figures. A topological property is that property
shared by all topologically equivalent figures. So,
one must first understand what is meant by topologically equivalent figures. A formal definition of
topological equivalence is not needed for our discussion. An intuitive (or heuristic) definition of
topological equivalence follows.

DEFINITION

Two figures are said to be topologically equivalent
if and only if one figure can be conceived of as
coinciding with another after the combination of
stretching and cutting and resewing described
below. We imagine that our figures are made of
rubber of unlimited elasticity and, in moving a
figure, we can stretch, twist, pull and bend it at
pleasure. However, we must be careful that distinct points in a figure remain distinct; we are not
allowed to force two different points to coalesce
into one point. We are allowed to cut such a rubber figure and twist it or loop it, provided that we
later sew up the cut exactly as it was before; that
is, so that points close together before we cut the
figure are close together after the cut is sewn.

When one figure has been converted into the other
by stretching and resewing cuts, we have shown
them to be topologically equivalent.

EXAMPLE:

The following objects model equivalent figures.
Each of the items in the following groups are commonly used as models of topologically equivalent
solids, surfaces, and curves. If you are not hampered
by the inaccuracy of the thick physical objects used
to model abstract sets of points, the following lists
will give you a better idea of topologically equivalent
figures.

a) equivalent surfaces:
top surface of a desk, top of a piece of paper lying
flat, surface of a staircase, wrinkled sheet of paper,
an army blanket, inside of a bowl.

b) equivalent surfaces:
empty tin can with its top soldered back on, tennis
ball, empty box, empty corked bottle, globe.

c) equivalent solids:
solid ball, grapefruit, a die, a rolling pin, a post.

d) equivalent curves:
the symbols U and I.

e) equivalent curves:
the symbols 2, 3, 5, and 7.

None of these five entities above is topologically
equivalent to any other. The five are topologically
distinct. Notice that a tennis ball and a solid ball
model things in different equivalence classes; the
tennis ball suggests a surface and the other ball, a
solid. Similarly, the top of a sheet of paper models
something topologically different from a tracing
pencilled around this sheet of paper, the former a
surface, the latter a curve. Physical objects, accom-

panied by appropriate explanations give excellent practice in topological differentiation. ⬥

Solid, surface, and curve are mathematical categories modeled by a solid ball, a tennis ball, and a rubber band, respectively. Analogous to the definitions made in Chapter 2, we mathematically define these three categories. Communication of these three definitions is made easier by using the term "circular disk" to refer to the set of points inside a circle.

DEFINITION

A *surface* is a set of points, each point having the property that a set of points immediately surrounding it is topologically equivalent to a circular disk. We will also allow a point of a surface to be on the boundary of such a disk. (In this way a surface can have an edge.)

A sphere and the outside of a hollow box are both surfaces by this definition, for you can visualize on each a circular disk molded to fit around any point of that surface, even around a corner point of the box. Careful reading of the definition will substantiate the mathematical habit of also naming as *surface* that which is composed of many disjoint pieces.

The union of a closed surface and its interior is a solid. But that is not a sufficiently general definition. We want a general definition of *solid* that is analogous to that of surface. In doing this, a solid turns out to be any set of points locally like a solid ball.

⬥ Note To Prospective Teacher:

Topology lends itself to creative class participation. Some relevant examples include both games and written assignments. One game, "Matching the Leader," is good for small groups in which all students are equipped alike with rubber bands, string, or pencil and paper. To match the leader in his group, each student must make a curve topologically equivalent to the one the leader holds up—yet not an exact replica.

A game suitable to prepared surfaces asks the student to match the model he has prepared with another model. The teacher can appoint certain people to go with their models to separate corners of the room, taking care that he chooses topologically distinct surfaces, one model to each station. The remainder of the class is instructed to mill about with their models. Each student settles at a station with a model he considers topologically equivalent to his or boldly stands at the center to proclaim his model different.

Naturally, the game can be expanded to include solids and a wide variety of models; but to be safe, a brief discussion of the usual materials and thicknesses representing surface should precede the game.

The teacher will find it worthwhile to endure the strain of waiting half an hour or until arguments cease before serving as referee, for the children will be most vocal in their attempts to oust an offending member who dared to go to their corner of the room with a surface inequivalent to the original surface.

Textbooks and workbook assignments are

DEFINITION

A *solid* is a set of points, each point having the property that a set of points immediately surrounding it is topologically equivalent to the interior of a sphere. We will also allow the point to be on the surface of such a solid ball. (In this way, a solid can have a boundary.)

Notice that this definition also calls a *solid* that which is made up of many disjoint pieces, each of which fits the previous definition.

DEFINITION

A *curve* is a set of points topologically equivalent to a line, a portion of a line (other than isolated points), or the union of any finite number of these. (The concept of line has to be taken here as undefined—intuitively understood; the formal definition is unbelievably complicated and not particularly elucidating.) The part of a line from some point on is called a *ray*. That portion included between two points is called a *line segment*, and it includes both end points.

Thus the definition of curve could be worded this way: a *curve* is a set of points topologically equivalent to a line, a ray, a line segment, or any finite collection of these three. (Note in Figure 8 the use of arrowheads to symbolize the direction of infinite extension.)

EXERCISE 2:

Make a pencil drawing symbolizing such things as lines and rays, as demonstrated in Figure 8. Start by placing two points, A and D, on a sheet of paper.

a)   Pencil in two rays that start at A but do not pass through D. Locate points B and C, one on each ray.

We call the rays $\overrightarrow{AB}$ and $\overrightarrow{AC}$.

b)   Pencil into the same drawing the line segments $\overline{AD}$ and $\overline{BD}$.

c)   Pencil in a line passing through the point A and containing none of the rays or line segments previously mentioned. Label so that this line is called $\overleftrightarrow{AE}$.

Look closely at the symbol for a line segment, and you will see that it is possible to remove two points without separating the segment into two disjoint parts. But a third point removed must separate the segment. It is also possible to remove one point (the beginning point) from a ray without separating it. We say that the line segment has two *noncut points*, and the ray, one. The line has no noncut points; all its points are *cut points*. For some curves, it is more definitive to state the number of cut points and let the reader infer (correctly) that all the others are noncut points.

FIGURE 9 Curve	No. of cut points	No. of noncut points
line		0
ray		1
line segment		2
simple closed curve	0	
figure-8	1	

Figure 9 shows cut points for the five distinct classes of curves most referred to in the next section. The number of cut points determines the separation

superfluous to play with topological equivalence. Write a letter of the alphabet on the board and you start a match or a "make different" game, a homework assignment to write all the equivalent letters and numerals, or an assignment to classify into equivalent sets all the capital letters of a printed alphabet. A sailors' knot or a string pattern, a twist of your body, or a rubber band can start other games going. Classroom demonstrations even include agile youths in sloppy sweatshirts; it is written that, theoretically, a man can remove his vest without removing his coat, a topological fact an agile youth can demonstrate with a stretchable or loose vest and a sloppy sweater serving as a dress coat.

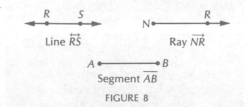

Line $\overleftrightarrow{RS}$

Ray $\overrightarrow{NR}$

Segment $\overline{AB}$

FIGURE 8

properties of a curve? The number of cut points and noncut points on a curve is preserved by an elastic motion, and so this characteristic of separation is a topological property. We will return to the topic of separation later to study separation of surfaces and curves.

EXERCISE 3:

Use the set of capital letters given here to practice separating these letters into classes according to the number of cut points. State to the right of each class the number of cut points or noncut points true for that class.

A B C D E F G H I J K L M N O P Q R S T U V W X Y Z

Showing two objects differ in one topological property suffices to show them inequivalent. And yet, showing that they are alike in one topological property is insufficient to show that they are equivalent, for they may differ in some other property. So the classes you have made in Exercise 3 are topologically distinct, but the counting of cut points is not sufficient to guarantee topological equivalence for all the members of each class. As an example of topological inequivalents, consider the curves of one cut point, the two-petaled flower we call the figure-8 and the four-petaled flower that looks like two figure-8's crossing at center.

Examples of curves abound in our classrooms, often as figures composed of line segments. Examples are such things as a square, a hexagon, and any polygon you might name. For that matter, the definition assures us that anything topologically equivalent to these is also called a *curve*. Because the nameless clod and the hexagon to its left in Figure 10 are both topologically equivalent to the square, they are also curves. Similarly, a spiral that continues outward indefinitely is a curve because it is topologically equivalent to a ray.

A skater's figure-8 is a curve since it is topologically equivalent to a collection of line segments such as two squares meeting at one point. One particularly important entity often overlooked in classifications is the angle; it is important to see it as a curve before embarking upon any considerations of measure. An *angle* is a union of two rays beginning at the same point, and hence it too is a curve.

We commonly call the rays that form an angle the *sides*, and call the end-point common to both rays the *vertex*. Often the letter at the vertex is used to name an angle, but sometimes three letters are needed to specify a particular angle. The angle in Figure 11 is rightly called angle A (written ∠A) or ∠BAT or ∠TAB.

EXERCISE 4:

a) Name all the angles you can see in Figure 12.

b) Figure 12 shows a pair of intersecting lines. State some difficulties that might arise if an angle were defined as "the geometric figure that is the union of two intersecting lines."

c) Suppose an angle were defined as the union of two line segments instead of two rays. This is a major switch from your previous concept of angle, affecting that which you draw. Give an example of

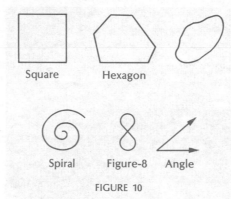

Square   Hexagon

Spiral   Figure-8   Angle

FIGURE 10

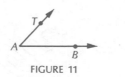

FIGURE 11

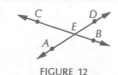

FIGURE 12

the disadvantage created in adopting such a definition, a creating of many angles where before there was one.

d) Let us strive to verbalize this discrepancy between the ray definition and the line segment definition. Hopefully, the following questions will allow you to articulate this, if you haven't done so previously. By the standard definition using rays, angle A in Figure 11 is the set of all points on the two rays which compose the angle. Is $\overline{AB} \cup \overline{AT}$ the same set of points? Is $\overline{AB} \cup \overline{AT}$ the same as the set of points $\overline{AC} \cup \overline{AN}$, where C and N are points beyond B and T on their respective rays? What happened to the unique angle when we changed to the line segment definition?

EXERCISE 5:

In each of the following, a boy is facing in one direction and then turns so that he is facing in another direction. Draw a figure for each exercise, indicating the angle through which he turns by placing an A at some point on the initial side.

a) From North to East

b) From East to West

c) From SW to SE

d) From E to SSE

e) From NNW to SW

a)
A

An angle is a curve topologically equivalent to a line. Our interest in angles will reawaken in Chapter 8, for that curve is of particular interest in the study of measure. However, to view angle from a topological standpoint, we are interested that it is a curve of infinite extent and "open." We classify curves in topology as open or closed. Now that we have named some specific curves of renown, we have sufficient vocabulary to discuss the topological properties of curves. A _closed curve_ is one having no loose end. A circle and all curves topologically equivalent to it are closed curves, as are complicated figure eights and skeletons of tetrahedrons and cubes. Those sets of points topologically equivalent to a circle are called _simple closed curves_, simple because the curve does not cross itself. The square and triangle and circle are all topologically equivalent and thus identical in their topological properties. For this reason, they are not differentiated in topology; they are all called "simple closed curves." We call angle, ray, and line segment open curves because they are not closed. An _open curve_ is any curve that is not a closed curve. ⦾

EXERCISE 6:

a) Label the curves in Figure 10 with the appropriate adjectives: "open," "closed," and "simple closed."

b) Label the curves in Figure 13 in the same way.

3.3 SEPARATION, A TOPOLOGICAL PROPERTY

Before we mentioned that a line is separated by a point, any point. An angle and all other curves topologically equivalent to a line have the same

⦾ Note to Prospective Teacher:
Here's an interesting competition game in which success is based on speed and perspicacity. Hand a group a sheet of paper with circled numbers 1–15 placed randomly on it. The object of the game is to start a pencil curve from the upper left corner of the paper, then connect consecutively numbered circles so that no curve crossing takes place, and finish in the upper right corner of the sheet. (Notice that the curve you must draw forms, with the top edge of the paper, a simple closed curve, since no crossing can take place.)

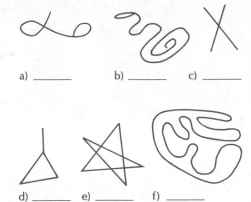

a) _____   b) _____   c) _____

d) _____   e) _____   f) _____

FIGURE 13

separation property. All curves have separation properties.

Surfaces and solids also have separation properties. Consider a simple closed curve drawn around a bug that is sitting on a sheet of paper. The bug (assumed a non-flier) cannot get from the inside to the outside without crossing the simple closed curve. We then say that the plane is _separated_ by a simple closed curve. Specifically, such a curve separates the rest of the plane into two sets of points. A line segment leaves the rest of the plane in one piece, and a figure–8 separates the plane into three sets of points.

Space can be separated, too—by certain solids and surfaces. (Not curves, as you will note if you hold in space a rubber band or a piece of string.) Space is separated by a surface if it is possible that a model of the surface could restrain a fly in its flight to another point in space. In general, a set of points, A, _separates_ another set, B, if there exists two points of B that cannot be connected without intersecting A. A sphere, modeled by a balloon, separates space, for no fly can get out or in without touching the surface. Space is separated by that figure modeled by an empty tin can with its lid soldered back on—and it is not separated by solids like those modeled by dense objects such as books, desks, and coffee cups.

EXERCISE 7:

a) State the number of pieces into which the plane is separated by each curve pictured in Figure 13.

b) State the number of pieces into which the plane is separated by each of the following curves: a simple closed curve, a line, a line segment, a ray, a figure-8, and an angle.

We can continue the discussion of the separation of surfaces by certain curves only if we coordinate our vocabulary so that we are all picturing the same surface. Below are the meanings of pertinent words.

The _plane_ is like the surface of an infinitely big desk or a surface that appears flat to a bug crawling along in any direction. (In Sadlier Series Contemporary Mathematics, Helen K. Halliday et al., Sadlier, N.Y., 1968, the student is asked to examine the surfaces of such physical objects as those listed in Example 1 and to classify each surface as plane or curved; you may want to try this yourself.)

The surface you make by rolling the sides of a sheet of paper together to make a tube is a piece of a cylinder. _Cylinder_ is the name for the surface of the infinitely long tube. (Notice that it feels flat to your touch if you run your finger in a certain direction. This is because the cylinder is like a skin stretched around a collection of lines.) By joining together the ends of a paper tube (piece of cylinder), you make a _torus_ or what is popularly called the donut or inner tube surface. The _sphere_ is the surface of a perfectly round ball; notice that, different from the others, the sphere does not feel flat anywhere as you run your finger away from a point on the surface.

EXERCISE 8:

Having settled on the meanings for these named surfaces, we can proceed with a check on what types of curves separate each surface.

a) State the number of pieces into which the cylinder is separated by a simple closed curve and the topological equivalent of each of the following: a line, a line segment, a ray, and a figure-8.

b) State the number of pieces into which the torus is separated by a simple closed curve.

Not only is the collecting of the separation properties of the different surfaces fun and educational in its own right; it also provides a check on which surfaces are topologically different. Something is awry if two surfaces we assumed topologically equivalent turn out to have different separation properties. The process of stretching will map any curve in one surface into a curve in the other surface. The new curve will be topologically equivalent to the old curve because the processes applied to the old surface will preserve the topological properties of the curve embedded in it. Furthermore, if the curve in the old surface separates the old surface, the curve in the new surface separates the new surface, and conversely.

If two surfaces are equivalent, then they are separated alike by each type of curve.

This observation can be used to show that certain pairs of surfaces are _inequivalent_. For example, the plane is not equivalent to the torus because every simple closed curve separates the plane but not the torus. (A torus is not separated by a simple closed curve going around the hole on the top surface as it lies on your desk, nor by one banding it.) Because of this difference in separation properties, the torus is also topologically different from the sphere and from the cylinder.

We mentioned that a balloon, modeling a sphere, separates space. So do the plane, the torus, and the cylinder. Closed surfaces capable of separating space into what we call an inside and an outside have two sides we could color. In playing with a piece of cylinder one inch high and at least 4 inches in diameter, we can see the obviousness of 2 sides to the cylinder surface. Yet, the same strip of inch-wide paper that formed the cylinder can be taped into what we call a *Mobius strip* by giving one end of the strip a half twist and taping the ends together as shown in Figure 14. As you run a pen along the "inside," you will discover that you have eventually marked the whole surface; there is only one side to the Mobius strip.

There is a surface called a *Klein bottle* that also has just one side. Because the Mobius strip and the Klein bottle have only one side, they are "nonorientable." Orientability is another topological property; that means that equivalent surfaces are all orientable—or are all nonorientable. You can find out what "orientable" means by reading Barr, pages 25–28.

By all means, see *Mathematics*, a volume of Life

FIGURE 14

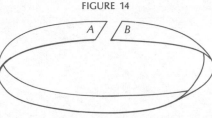

Tape *A* and *B* Together

Science Library, for excellent pictures and a brief discussion of some topological properties of Mobius strips and Klein bottles. Those surfaces are mind-expanding to explore, for they stretch our imagination and our conception of space and orientation. The *Klein bottle*, besides not separating space and being nonorientable, contributes to our notion of a fourth spatial dimension. It is a closed surface made from a long piece of cylinder in a way analogous to that by which we constructed a Mobius strip from a strip of paper. The ends are attached after one end approaches the other from inside, something that is impossible to see without visualization of a fourth dimension, for in 3-space it appears there is an intersection; yet it only appears that way, just as some photos of a rubber band held in a near figure-8 make it appear (in 2-space) that the rubber band intersects itself. (To understand the nature of 4-space, read a good satire by Abbott on the people of limited viewpoint who inhabit Flatland.)

In this chapter, we discussed topological properties of equivalent figures and the meaning of topological equivalence. Frequently you can use what you have learned here to show that two figures are topologically the same or topologically different. If two figures are topologically equivalent, we *may* be able to convince someone of this fact by demonstrating a cutting and deforming that transforms one of the figures into the other. For example, given a rubbery sphere with a hole in it, we could pull on the edges of the hole until the hole was so large that all the surface could be stretched into a piece of plane. (No cutting and resewing are needed.) This would perhaps convince one that a sphere with one section of its surface missing is really topologically the same as a piece of a plane.

Proving two figures are topologically inequivalent can only be done by showing that they differ in one of their topological properties. We can illustrate this procedure by proving that a cylinder is not topologically equivalent to a plane.

PROOF

There are two kinds of positionings of a line on a cylinder—the straight line running the length of the cylinder as one of its ribs and its equivalent spiraling the cylinder like an endless coil. Neither of these separates the cylinder. A plane is separated by a line. Hence, the cylinder is topologically inequivalent to the plane.

This chapter has introduced you to topological equivalence and topological properties and has simultaneously enriched your vocabulary of various surfaces and curves. Understanding topological equivalence and this modest vocabulary of topological terms is necessary to Chapter 5 and 6 and basic to the rest of this text. Having this topological background will enable you to take a completely different approach to geometry of curves, surfaces, and solids. You can enter the world of networks (curves) armed with a rubber band and survive as gloriously as did David with his slingshot; from there you can step lithely into the middle world of maps (surfaces) and feel the rhythm of the whole of geometry.

ANSWERS TO EXERCISES

1. *c* and *d* are equivalent to Figures 1–5.

3. With them printed as I have here, one class would look like this:

C I J L M N S U V W Z

This is the class of letters having two noncut points (and all other points are cut points).

4. a)   ∠CED   ∠BED   ∠BEA   ∠AEC   ∠BEC   ∠AED.

   b) Figure 12 would be just one angle, confounding those interested in the measure of an angle.

   d) No, not the same set of points as ∠A was defined in Figure 11.
   No.

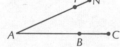

If an angle were the union of two line segments, then ∠BAT would be different from ∠CAN in this picture. Previous definition made them the same angle.

5. a)   b)   c)

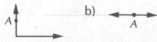

   d)   e)

6. a)   a)   closed.   b)   closed.
         simple closed.   simple closed.

   c)   closed.   d)   open.
        simple closed.

   e)   closed.   f)   open.

6. b)   open   open   open
        open   closed   closed
                  simple closed

7. a) 3   b) 2   c) 1   d) 2   e) 7   f) 2

7. b)   simple closed curve: 2
        line: 2
        line segment: 1
        ray: 1
        figure eight: 3
        angle: 2

8. a)   simple closed curve: 2
        line: 1
        line segment: 1
        ray: 1
        figure eight: 3

   b) simple closed curve either separates the torus into two pieces or it doesn't separate it at all.

Abbott, E. A. *Flatland: A Romance in Many Dimensions*. 5th rev. ed. New York: Barnes and Noble Books, 1963.

Arnold, B. H. *Intuitive Concepts in Elementary Topology*. Englewood Cliffs, N.J.: Prentice-Hall, Inc., 1962.

Barr, Stephen. *Experiments in Topology*. New York: Thomas Y. Crowell Co., 1964.

Bergamini, David. *Mathematics*. Life Science Library. New York: Time-Life Books, 1970.

Courant, Richard, and Robbins, Herbert. *What Is Mathematics?* New York: Oxford University Press, 1960, pp. 235–67.

"Topology." *The World of Mathematics*. Vol. 1. Edited by James R. Newman. New York: Simon & Schuster, Inc., 1956, pp. 581–99.

Denholm, Richard A. *Mathematics: Man's Key to Progress*. Pasadena, Calif.: Franklin Publications, Inc., 1968, pp. 75–81.

Froman, Robert. *Rubberbands, Baseballs and Doughnuts: A Book About Topology*. New York: Thomas Y. Crowell Co., 1972.

Gardner, Martin. "Knotting Problems With a Two-Hole Torus." *Scientific American* (December 1972), p. 102.

Garstens, Helen L., and Jackson, Stanley B. *Mathematics for Elementary School Teachers*. New York: The Macmillan Co., 1967, pp. 136–7.

Halliday, Helen K. *Sadlier Series Contemporary Mathematics*. Sadlier, N.Y., 1968.

Hartung, Maurice L., and Walch, Ray. *Geometry for Elementary Teachers*. Glenview, Ill.: Scott-Foresman & Co., 1970, pp. 1–19, 31–37.

Rosenthal, Evelyn B. *Understanding the New Mathematics*.

Greenwich, Conn.: Fawcett Publications, Inc., 1966, pp. 217–27.

Rosskopf, Myron F.; Levine, Joan L.; and Vogeli, Bruce R. *Geometry: A Perspective View*. New York: McGraw-Hill, Inc., 1969, pp. 256–63.

Sharp, Evelyn. *A Parent's Guide to the New Mathematics*. New York: Simon & Schuster, Inc., Pocket Books, Inc., 1964.

Tucker, Albert S., and Bailey, Herbert W., Jr. "Topology." *Scientific American* (January 1950).

Van Engen, Henry; Hartung, Maurice L.; and Stochl, James E. *Foundations of Elementary School Arithmetic*. Glenview, Ill.: Scott-Foresman & Co., 1965, pp. 379–93.

tessellations on the plane

MATERIALS NEEDED:

Cardboard or construction paper, tracing paper, scissors, four crayons of different colors

Suggested Classroom Presentation:

Silent demonstration on overhead following Exercise 5

◑ Note to Prospective Teacher:
The Nuffield Mathematics Project in Great Britain made play with tessellations popular with many British children, and the booklets and speakers from that group have assured its popularity in the elementary schools of the USA.

In this play chapter, we will be exploring tessellations on the plane. A *tessellation* is a pattern of regions used to cover a surface without overlapping regions or leaving gaps. Playing with these patterns of regions in the plane, to try to fill the plane, has become a popular mathematical game. Play with tessellations develops intuition regarding the plane and components of surfaces in general. Above all, play with tessellation leads to an understanding of area. ◑

We see tessellation patterns in mosaics, in ceiling and floor tiles, and in sides of constructed walls. All the patterns in Figure 1 were found underfoot around a school.

In flooring, wide use is made of triangular patches composed of half of a square region. Must the triangles be half squares—their longest sides being the square's diagonal—in order to make a tessellation pattern? The faces of brick and cement block walls are great examples of tessellation patterns that could be continued to fill a plane.

Flemish Bond

Front View

Stretcher Bond

FIGURE 2

Floor Patterns

Paving Stones

FIGURE 1

FIGURE 3

Many tessellation patterns appear ornate. One can make a basic element for an interesting tessellation just by positioning some squares together to make a region.

EXERCISE 1:

Note in Figure 3 two possible designs that result from overlapping congruent square regions. Experiment by making three of your own designs by tracing around a square in different positions. Show at least one design that could serve as the basic element in a tessellation pattern, indicating how it fills the plane by displaying duplicates of the design in alternating black and white.

Tessellations need not use only one shape. The definition of a tessellation includes any shape, and thus a set of shapes or a pattern of shapes. Patterns of patches used to cover a plane surface without gaps or overlapping, i.e., which fill the plane, are called _tessellations_ on the plane. The pattern or set used repeatedly in the same position is the _basic element_ in the tessellation.

A flowing tessellation pattern drawn by Escher[1] is reproduced in Figure 4. This man astride his horse is used solely in the tessellation pattern, though the basic element in the pattern is an adjoining white and gray pair. The basic element, the pair in this case, is the smallest you could use as a printing block if you wanted to run paper past in vertical and horizontal directions.

Symmetry Drawing A

FIGURE 4

Symmetry Drawing B

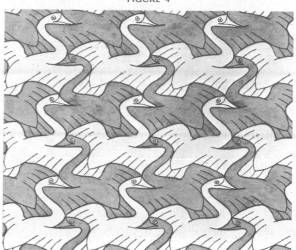

---

[1]M. C. Escher, _The Graphic Work of M. C. Escher_ (New York: Ballantine Books, Inc., 1971), p. 10, by permission of the publisher

FIGURE 5

Because of the technological process of printing, both textiles and wallpaper furnish numerous examples of tessellation patterns. Even though the basic element we search for in the tessellation pattern is usually smaller than the region used in actual printing, displays of wallpaper and cloth furnish practice in discerning the basic elements of a tessellation.

Test yourself on spotting the basic element in the proposed tessellation in Figure 5a.

I perceived Figure 5a as a tessellation of one shape that I then cut out and traced repeatedly to make sure that the one shape would indeed cover the plane. Here is a replica of my construction paper patch, with relevant comments.

Repeated element:

COMMENTS

It does cover the plane

1. There is a single shape repeated.

2. This shape is repeated over and over again as it moves to the right and downwards.

3. It is used only in this position.

4. Note the similarity of this shape to square regions that also compose a tessellation.

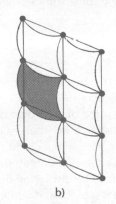

a)

b)

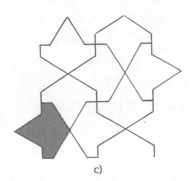

c)

d)

From "A Tiny Treasury of Tessellations," by Ernest R. Ranucci, *The Mathematics Teacher*, (February 1968), p. 115, used with permission. Copyright by the National Council of Teachers of Mathematics.

Figure 5 Continued

MATERIALS NEEDED ① 47

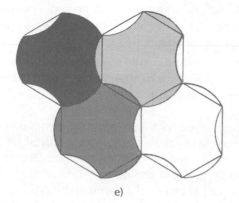

e)

f)

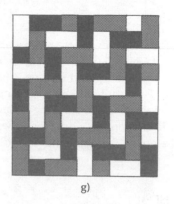

g)

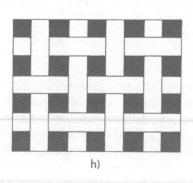

h)

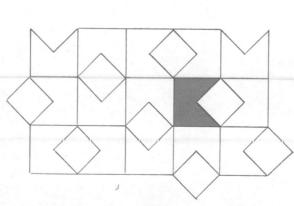

i)

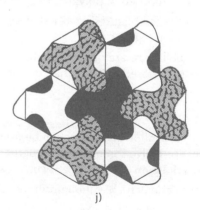

j)

EXERCISE 2:

In *The Mathematics Teacher*, February 1968, Ranucci published Figure 5, claiming that each picture suggested a tessellation of the plane. Check Ranucci's claim, while identifying the basic element of each proposed tessellation. For each tessellation, present this basic element (in a drawing or with pasted paper) along with comments, as I did for Figure 5a.

EXERCISE 3:

Which, if any of the tessellations you saw in Ranucci's figure are composed of polygonal regions?

Which of the ten pictures are tessellations of just one polygonal region?

Which tessellations are of a region that appears complex (composed of more than one shape)?

EXERCISE 4:

Which of the following shapes can be used to make a tessellation? Give some rationale, perhaps a drawing, to back up an affirmative answer.

a)   a circular patch? b)   the imprint of your foot?

c)   the cover of a   d)
book?

e)

EXERCISE 5:

Draw any big triangle on heavy paper. Cut out the

triangular patch. Make duplicates of this triangular patch and see if you can fit these together to make a tile pattern so that there are no gaps and no overlap. In other words, see if you can make a tessellation on the plane using only that triangular patch.

State your opinion as to the possibility of tessellating the plane with a randomly chosen triangle, giving a reason, if possible.

The questions relating to tessellating with polygonal regions can each be answered scientifically, but this approach is not immediately obvious. Demonstrations on the floor and on an overhead projector would be of great help to you at this point. This is because observation of successful tessellation patterns and movement suggesting properties of the polygon under consideration aids in deciding whether or not a tessellation is possible. The next exercise helps narrow your attention to these considerations.

EXERCISE 6:

Some interesting regions to explore are those bounded by rectangles, weird quadrilaterals, regular pentagons, regular hexagons, and regular octagons. (A *regular polygon* is one composed of segments of equal length.) If a regular hexagon of 12-inch sides will work, then one larger but shaped the same will work. The same reasoning applies to any polygon. Divide up the topics so that each individual works on one of the last four suggested here.

a)   Organize a silent demonstration for your chosen

region. You will find that numbering or coloring each angle of the region and purposefully moving your regions on the floor or on an overhead projector will allow others watching to see that your region will or will not serve as a single element in a tessellation of the plane—and this demonstration well done, will help them understand why. (See parts c, d, and e.)

b) Using sketches and words, write a short explanation as to *why* your region will or will not serve as a single element in a tessellation of the plane. (See parts c, d, and e.)

c) If we call the total angle measure in a triangle 180 degrees, how many degrees would we say are

in a rectangle?
in a general quadrilateral?
in a regular pentagon?
(Remember, a regular pentagon has 5 equal sides, all line segments.)

in a regular hexagon?
in a regular octagon?

Have the nerve to say "don't know" if you think something is indeterminable.

d) If there are 180 degrees of an angle measure in a triangle, how many degrees are there

in a single angle of a rectangle?
in a single angle of a regular pentagon?
in a single angle of a regular hexagon?
in a single angle of a regular octagon?

e) How many degrees does the second hand of a clock sweep through in one minute? If patches can be fit together in a tessellation of the plane, then what is the total number of degrees around any point in the tessellation pattern?

Finding a tessellation must not be confused with finding area. When finding area, we concern ourselves with covering a surface enclosed by a boundary and then measuring it in some way; when considering tessellations, we shall not consider any boundary, i.e., we could go on and on making the pattern. Also, the concept of area is traditionally the counting of tessellated squares needed to fill up this bounded region. When we study tessellations, we study many regions besides square regions, each of which could be used—as the square is—in the measure of area.

ANSWERS TO EXERCISES

3. The tessellation patterns labeled *c*, *d*, *g*, *h*, and *i* are tessellations of polygonal regions.

   Figures 5*c*, *d*, and *g* are tessellations of one polygonal region.

   Two of the patterns appear to be tessellations of complex regions.

4. Only patches labeled *c*, *d*, and *e* can be used as the basic element in a tessellation pattern. Four duplicates slid together will be sufficient to assure you of this fact.

6. c) A quadrilateral is the union of two triangular regions so there are twice 180 degrees (360 degrees) in the angles within the quadrilateral, thus a rectangle. Similar reasoning shows $3 \cdot 180°$ in a polygon, and so forth for the other polygons.

   d) A rectangle is a quadrilateral with all its angles of equal measure. Hence, each angle is $\frac{1}{4}(360)$ degrees. Whenever a polygon discussed is regular, all its angles are of equal measure and so each can be computed using the information just computed in 6 *c*.

   e) 360°. The total angle measure of all the angles meeting at a point in a tessellation must be 360 degrees.

BIBLIOGRAPHY

Eves, Howard A. *A Survey of Geometry*. Vol. 1. Boston: Allyn & Bacon, Inc., 1963, p. 428.

Nuffield Mathematics Project. *Shape and Size 3*. New York: John Wiley & Sons, Inc., 1968, pp. 24–26, 38.

Ranucci, Ernest R. "A Tiny Treasury of Tessellations." *The Mathematics Teacher* (February 1968), pp. 114–16.

Van Engen, Henry; Hartung, Maurice L.; and Stochl, James E. *Foundations of Elementary School Arithmetic*. Glenview, Ill. Scott-Foresman & Co., 1965, pp. 374–9.

chapter 5

networks

MATERIALS NEEDED:

Pipe cleaners
Skeleton of a tetrahedron
Optional:
Printed circuit board printed only on one side

Now we begin a study of networks. A curve thought of in a special way is called a network. Networks have some topological properties in common, some of which have particular application. You will see, in this chapter, that topological equivalence of a town's bridges to a certain network allows one to schematize uniquely the routes available in the town. In this way, the "size" and quantity of all junctions is plainly displayed. Also, all equivalent networks have the same separation properties. Therefore, in Chapter 6 we see that all equivalent networks split a plane map into the same number of sections. In addition to learning what a network is and how to draw one applicable to a problem, you will explore the traversability of a network and learn of the polygons as a special network.

5.1 THE NETWORK AS A SCHEMA

Before beginning the study of networks, it is important that you recall certain vocabulary that you studied in Chapter 3. The following review should be sufficient. Next to each phrase below indicate

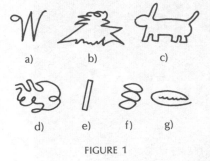

a)   b)   c)

d)   e)   f)   g)

FIGURE 1

each curve of Figure 1 that fits that description.

open curve _____

simple open curve _____

closed curve _____

simple closed curve _____

The only simple closed curves are those in c and e of Figure 1. Figures a and d are nonsimple open curves, with a simple open curve being displayed only in g.

A simple curve, one piece of curve uninterrupted, can be referred to as a curve of one arc. One arc or a union of arcs is called a _network_.[1] This term was borrowed from the people in fields such as transportation and communication. The language of networks offers to mathematicians a vocabulary in which to conceptualize certain basic properties of curves—topological properties.

Many puzzles and mathematical teases are based on networks, as are schemas for such things as games, wars and efficient production. When you get warmed up, see "Applications of Networks" at the end of the chapter for some of the latter. As an example of the former, consider the classical question of moves of a knight on a chessboard.

---

[1]This configuration can also be referred to as a "graph." This term is more modern, but has many meanings, and so is avoided here. Although I shall use "network," the student who wishes to do further reading should be aware that the term "graph" is common in even very rudimentary treatments.

An exercise for chess players: Is it possible to move the knight from some arbitrary starting position around the whole board and return it to the starting point so that each square has been occupied just once? (Even the nonchess player can see that you will have a closed curve composed of a union of arcs.) If each arc represented a move, how many arcs does any solution have?

As you view a few examples of networks, you will realize that they are used by children, many adults including electricians, and thoughtful hosts whose guests need a rough map. The railway company serving greater Copenhagen publishes the network in Figure 2a as part of a schematic diagram of their local train service. The arcs are those lines between the junctions or intersections. Topologically this diagram is the same as the curving mass of tracks appearing on a map of Copenhagen (Figure 2b). Both map-reader and the diagram-reader will reach their destination; the diagram-reader, however, will spend less time figuring out the necessary connections. The wiggles in the route and the length of the route are negligible factors in travelling by local trains in this compact city. The places of transfer and the number of stops are factors that would make a tremendous difference in trip time, and these the schematic diagram can preserve.

Networks are printed or sketched by many companies and individuals as a way of isolating the relevant facts. Both Figures 2a and 2b were easier to grasp than the city map with its confusing maze of streets and tracks. In the same way an airline

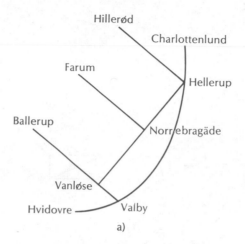

a)

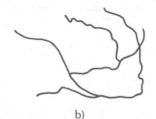

b)

FIGURE 2

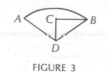

FIGURE 3

diagram schematizing cities with daily connections is easier reading for a prospective customer. Figure 3 is a copy of an airline's poster displaying the extent of its daily service between the four cities, A, B, C, and D.

You now have an idea of what a network will look like. Each open curve you see in the train diagram (Figure 2a) is called an _arc_, and the arcs are connected only at their end points. You can reorganize your concept of the train diagram so that you perceive a union of arcs. Counting them in Figure 2a, you should find nine arcs. A _network_ is defined as a figure composed of a finite, nonzero, number of arcs, each two of which intersect at their end points. The end points of these arcs are called the _vertices_ of the network. Each arc has two vertices, while a point in space can serve as a vertex for many arcs. You can count nine vertices in the train diagram, those representing the places of origin and transfer. In Figure 3 you see a network of four vertices and four arcs.

For purposes of communication, we speak of the _degree of a vertex_ as the number of arcs meeting there. In Figure 2, there are five vertices of degree 1, three of degree 3, and a vertex of degree 4 at Hellerup. In Figure 3 the degree of both B and D is 3, and A and C are considered of degree 2 if thought of as vertices.

EXERCISE 1:

State whether or not each part of Figure 4 is a network. If yes, state the number of arcs and the number of vertices. (Omit counting any vertices of order 2, for their presence is debatable.)

EXERCISE 2:

Figure 5a shows the possible routes between Manhattan and New Jersey, and Figure 5b shows the network of these routes. Note that the bridged pieces of land become vertices in the network diagram. This is in accord with the intent of the diagram: to emphasize the available routes between Manhattan and New Jersey.

a)   Sketch as a network the available routes of travel between the inside and the outside of a one-room cabin with two doors. Consider inside as one place and outside as another place.

b)   Sketch, as a network, the available routes of travel between the five rooms and yard of the luxurious beach house in Figure 6.

Is it possible to walk through each door exactly once?

c)   Sketch as a network the available routes of travel connecting the four parts of the town of Konigsberg, built on the confluence of the New Pregel and Old Pregal rivers in East Germany. Figure 7 shows a sketch of the old town with its seven bridges.

Is it possible to walk through the whole town without retracing your steps over any bridge? (This does not necessarily mean that you have to end at the starting point.)

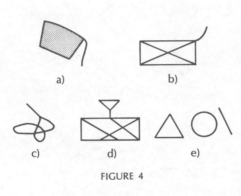

a)   b)

c)   d)   e)

FIGURE 4

New
Jersey   Manhattan

a)

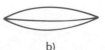

b)

FIGURE 5

5.2 TRAVERSING

Looking at your sketch of the network of bridges should make it easier to tell whether one can or cannot tour all of Konigsberg in a continuous walk without recrossing any bridge. We call that complete tour the _traversing_ of a network. The _traversing_ of a network consists of tracing over all arcs of the network without tracing over any arc twice. A great man named Euler devised the idea of sketching the town as a network and observing its properties in order to settle the question of the walking tour, much debated over beer steins.

Leonard Euler (1707–1783), was the most eminent mathematician of the eighteenth century. Born in Basel, Switzerland, he studied mathematics there under Johann Bernoulli. At age 20, he accepted the chair of mathematics at the new St. Petersburg Academy formed by Peter the Great. Euler was extremely well-read and a voluminous writer on mathematics. The most prolific writer in the history of science, he enriched mathematics in almost every branch of the study. His energy was at least as remarkable as his genius. "Euler calculated without apparent effort, as men breathe, or as eagles sustain themselves in the wind."[2] He worked with such facility and speed that it is rumored that he wrote mathematics papers while grandchildren swarmed the room and dashed off his memoirs

---

[2] James R. Newman, ed., "Commentary on a Famous Problem," _The World of Mathematics_, Vol. 1 (New York: Simon & Schuster, Inc., 1956), p. 571.

in the half-hour before dinner. He produced 13 children and sufficient mathematical papers (in at least 4 languages) to fill 79 enormous volumes, recently published as *Opera Omnia*.[3] It is of interest to note that his amazing productivity was at least as prodigious after 1768 when he became blind. Euler, after discovering a scientific principle concealed in the Konigsberg problem, presented his simple and ingenious solution to the Russian Academy at St. Petersburg in 1735. We translate here for you the beginning of this paper, showing you Euler's approach to the problem. An exciting film presentation of the problem,[4] other English translations,[5] and biographical notes on Euler[6] are readily available.

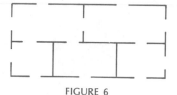

FIGURE 6

FIGURE 7

[3] Leonard Euler, *Opera Omnia*, ed. Ferdinand Rudio, Adolf Krazer, and Paul Stackel, ser. 1, vol. 17 (Leipzig: B. G. Teubneri, 1911–62), pp. 1–10.

[4] Bruce and Katherine Cornwell, "The Seven Bridges of Konigsberg," International Film Bureau, 1965.

[5] Leonard Euler, "The Seven Bridges of Konigsberg," *The World of Mathematics*, vol. 1, ed. James R. Newman (New York: Simon & Schuster, Inc., 1956), pp. 573–80; idem, "The Solution of a Problem Belonging to the Geometria Situs," *A Source Book in Mathematics: Twelve Hundred to Eighteen Hundred*, ed. D. J. Struik (Cambridge: Harvard University Press, 1969).

[6] Eric Temple Bell, *Men of Mathematics* (New York: Simon & Schuster, Inc., 1961); Turnbull, Herbert W., "The Bernoullis and Euler," The World of Mathematics, vol. 1, ed. James R. Newman (New York: Simon & Schuster, 1956), pp. 148–51.

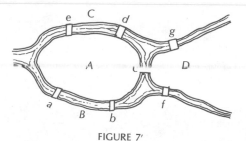

FIGURE 7'

The Konigsberg Bridge Problem
L. Euler

1. Besides that part of geometry that deals with quantities and has been studied intensively at all times, there exists another one almost unknown up to now, which Leibniz mentioned first, calling it the geometry of position. This branch of geometry deals with all that which can be determined by position and with the investigation of those perperties that pertain to position alone; it does not take magnitudes into consideration, nor involve computation with quantities. But what kind of problems should be included in this geometry of position and what method shall be utilized for their solution has not yet been sufficiently determined. Recently there was a problem mentioned that apparently belonged to geometry, but was formulated in such a way that it neither required the determination of a quantity nor could it be solved by quantitative calculation; consequently, I did not hesitate to assign it to the geometry of position, especially since its solution required only the consideration of position, computation offering no benefit whatsoever. Therefore, I shall discuss the method that I developed for the solution of such problems, to serve here as an example of the geometry of position.

2. The problem, supposedly quite well known, was as follows: In the town of Konigsberg in Prussia, there is an island, *A*, called "the Kneiphof," with two branches of the river (Pregel) flowing around it, as shown in Figure 7'. Over the branches of this river lead seven bridges, *a*, *b*, *c*, *d*, e. *f*, and *g*. Now the question is whether somebody could plan a walk so as to cross all bridges once but not more than once. I have been told that some deny this possibility, others express doubt, but that nobody so far has produced

a solution. Based on this, I formulated for myself the following, most general problem: Whatever the shape of the river and the distribution of its branches, and whatever the number of bridges, to determine whether or not it is possible to cross each bridge exactly once.

3. The particular problem of the seven bridges could be solved by a careful enumeration of all possible walks, thereby ascertaining which, if any, satisfies these conditions. This method, however, is much too tedious and difficult because of the great number of combinations, and it could not possibly be applied in cases where a great many more bridges are involved. If the investigation were conducted in this manner, much would be found that was not asked for; no doubt this is the reason why this approach would be so taxing. That is why I have dropped this method and looked for another, leading only so far as to show whether or not such a walk can be found; such an approach, I believed, would be much simpler.

4. My whole method is based on an appropriate and convenient designation of the crossing of the bridges. To do this, I use the capital letters, *A*, *B*, *C*, and *D*, to name the individual regions separated by the river. Then if one reaches region *B* from region *A* by crossing either bridge *a* or bridge *b*, I denote this crossing by the letters *AB*, the first of which names the area from which the traveler emerges, while the second indicates the area reached after the crossing. If the traveler then goes from the region *B* over bridge *f* into *D*, this crossing will be designated *BD*; those two consecutive crossings, *AB* and *BD*, I shall simply designate by three letters *ABD*, where the middle letter indicates the area into which the

first crossing leads as well as the area out of which the second crossing leads.

5. ... Hence, the crossing of seven bridges will require eight letters for its description.

Euler's solution continues in this very readable manner, but we shall cut it here so that you can find for yourself the joy of discovery Euler experienced.

An organized examination of some other networks while checking for traversability will likely lead you to discover what Euler saw. Throughout the following exercises, keep a notebook with your observations about traversability of networks. Its headings will include the degrees of the various vertices.

There is one note that must be injected before you start counting vertices of each order; because of the fact that an arc can manifest multitudinous corners and bends, vertices of degree 2 are senseless to count. Remember, the *degree of a vertex* is the number of arcs meeting at that point. You can see none or see an infinite number of these second degree vertices on a simple closed curve. Take heart, for there is a convention allowing us to communicate: we consider a simple closed curve as having one vertex. Thus, Figure 8a can be thought of as one arc and one vertex, Figure 8b as two arcs and two vertices, and Figure 8c as four arcs and four vertices. This manner of treating vertices of degree 2 is merely a convention; considering more such vertices present on any arc will not affect any theorems or practical applications of networks.

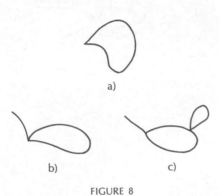

a)

b)   c)

FIGURE 8

EXERCISE 3:

Use the definition of network given in this chapter to answer the following. Draw networks to substantiate affirmative answers.

a) Is there a network with 50 arcs and 1 vertex?

b) Is there a network with 1 arc and 50 vertices?

c) Draw a network with 5 arcs and 8 vertices. Are all others fitting this description merely topological equivalents to the one you just drew? If not, draw another.

Back to traversing. Head a notebook sheet with sufficient information to designate the network, the number of paths needed to traverse the network, and a count of the degree of the vertices and number of arcs so that you can see a relationship as Euler did. Obviously, we need not list the number of vertices of degree 2. Following is a partial heading leaving you room to fit in other columns you find helpful.

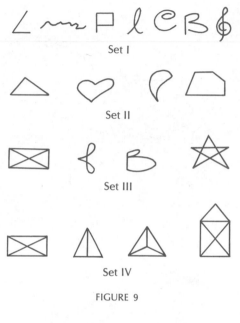

Set I

Set II

Set III

Set IV

FIGURE 9

Tracing a whole network without lifting the pencil is a physical expression of the act mathematicians call *traversing* the network in one path. If the pencil must be lifted for you to trace the network, then more than one path is needed to traverse the network. Exercise 4 further amplifies this and gives you experience in tracing as a physical analogy.

EXERCISE 4:

a) Without lifting your pencil from the plane of your paper and without retracing an arc, try to copy each of the drawings in Figure 9 in your notebook. A simple arc such as are the first few in Set I can be traced by putting the pencil down just once. Record beneath your tracing the total number of times you found necessary to put the pencil down. This is the number of paths needed to traverse the network.

b) The curves in Set I are all open curves, and they can all be traced in one path.
Can all open curves be traced in one path? Can all simple open curves?

c) The curves of Set II and Set III can all be traced in one path. Can all closed curves be traced in one path? Can all simple closed curves?

Notice with the open curves that a limb sticking out, like on the capital letter *P*, either has to be a beginning spot or an ending spot if you succeed in tracing the curve in one path. Compare this to the capital letter *A*.

A *path* in a network is a sequence of different arcs

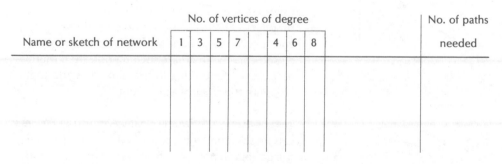

Name or sketch of network	No. of vertices of degree							No. of paths needed
	1	3	5	7	4	6	8	

in the network that can be traversed continuously without retracing any arc. The curves in Sets I, II, III, IVb, and IVd are all networks that can be traversed in one path. The networks pictured in IVa and IVc can be traversed in two disjoint paths, and not less than two suffice.

EXERCISE 5:

How many paths are needed to traverse Copenhagen's train system and each of the networks of Exercise 1.

EXERCISE 6:

a)   Traverse each of the networks in Figure 10. Use as many paths as are needed. Record the vertices in the order in which you pass through them, putting all vertices of a path inside a parenthesis as has been done for Figure 10a.

In Figure 10a, vertices A and F are of degree 1, vertices B and C are of degree 3, and the others are of degree 2. The four vertices—A, F, B, and C—are called odd vertices. A vertex of odd degree is said to be an *odd vertex*.

Writing the sequence of vertices according to their degrees may lead you to make a connection between the degrees of the various vertices you pass through and the number of paths needed to traverse a network. Figure 10e can be traversed in as few as two paths. Here is one possibility: (LGEMBADCDCBAC) (DMEFFGHIKIHLK). Notice the respect-

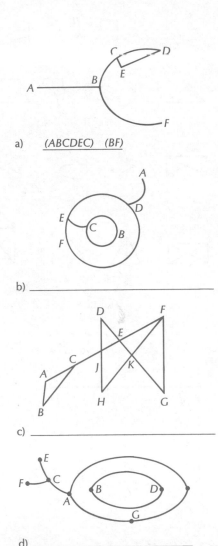

a)   (ABCDEC) (BF)

b) _____

c) _____

d) _____

FIGURE 10

ive degrees of the vertices on each path in this way of tracing Figure 10e: (3,4,4,4,4,4,5,5,5,5,4,4,5) (5,4,4, 4,4,4,4,3,4,4,3,3). Now do the same for your traversing of e if your paths were different from the above. Invent two other paths that traverse the network of 10e and compare the numerical sequences denoting the degrees of the vertices.

EXERCISE 6: (continued)

b)   Write similar numerical sequences as we did for 10e beneath each letter sequence of Figure 10. Compare notes with each fellow student whose paths begin differently from yours; enter, near the figures, other paths and the numerical sequence associated with each traversing.

c)   Draw any conclusion you can about the beginning and ending vertices of the various paths composing the efficient traversing of a given network.

Because we are studying topological entities when we are studying networks, we should waste no time examining myriad networks that are really topologically the same. If we have examined one simple closed curve, there's no point in examining another. And if a single path traverses a simple closed curve with an arc sticking out, then there's no point in continuing to examine networks that are mere topological equivalents.

Two networks are topologically equivalent if one can be stretched, etc., to form the other. Reread the definition of topological equivalence in Chapter 3. From the definition it is obvious that the two sketches in Figure 11 are equivalent.

And, as you would expect, the degree of each vertex is maintained in the transformation from the first figure to the second. An interesting question is whether any two networks are equivalent if they have the same degree at respective vertices. For example, will all networks having four vertices of degree 3 turn out under investigation to be equivalent to Figure 11? For further work on this question, see Exercises 14–18.

Several short paragraphs follow, each designed to help you conclude, by looking at a network, how many paths are needed to traverse that network and all networks topologically equivalent to it. This was the problem that Euler simplified and solved when he settled the bets of the citizens of Konigsberg. It is sufficient for you to concentrate on finishing the sentence, "If a network is traversed by a single path, then. . . ." (What you are looking for in the end is a description of the appearance of the network such as the number and type of vertices or arcs.) The following paragraphs are each designed to give you leads on those networks which can, and those which cannot, be traversed in one path.

EXERCISE 7:

a)   Make a chart on a full sheet of paper, concentrating first on listing networks of topological types that can be traversed in one path. Suggested column headings for the chart are given below. Notice as much as you can about classes of networks. As a starter ask yourself, "Are there any types of networks I can traverse in one path that begin and end at the same vertex?" (There are.)

Name or sketch of network	No. of vertices of				Total number of odd vertices	Total even vertices	No. of paths needed
	degree 1	degree 3	degree 4	degree 5			
Set I a,b, e, f, g, Set II	2				2	*all*	1 1

b)   Even if some observations are not profound, complete the following sentence as many ways as you can: "All networks that . . . can be traversed in one path." As each sentence comes to you, add it to this list. Obviously, only connected networks have a chance of being traversed in one path, so let's consider only those in which no arc is disjoint from all other arcs.

Notice all figures in Set II (Fig. 9) are topologically equivalent to a simple closed curve; hence they are all the same network having one arc and one vertex of order 2. Also, many of the figures in Set I are topologically equivalent to the second network in that set,—a network of two vertices. But these are the simplest networks of all those with one and two vertices. Figure 12 has some other networks. To form general opinions on networks of one vertex, examine the top line of Figure 12 below, and imagine or sketch an increasingly complicated series of networks of one vertex. Similarly, expand logically the patterns for networks of two vertices.

Notice that the vertices of degree 2 are those you must pass through because there is an arc leading in and an arc leading out. If you begin at such a vertex, you will end there and close your path. So, for

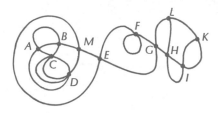

Figure 10   Continued

FIGURE 11

practical purposes, you can be said to have passed through. Henceforth, we will include a vertex at which we begin and end in the broad category of those vertices we must pass through.

EXERCISE 7':

a) Are vertices of degree 2 the only vertices that must be passed through?

b) Sketch three additional networks for each line of Figure 12, noticing the sequence of the vertices as you trace each network. For instance, to traverse the capital letter P in one path, one way is to start with the vertex of degree 1, pass through those of degree 3, 2, 2, 2, and finally finish at the vertex of degree 3. You may have traced it by starting at the vertex of degree 3, then on to 2, 2, 2, 3, and 1. That's just the reverse of the first way. If you traverse capital P in one path, you cannot start at any of the vertices of degree 2. Similarly, the only single-path traversing the second picture in Set IV of Figure 9 begins at the center pole of the tent; either way it involves tracing a sequence of vertices of degrees 3, 2, 3, 3, 2, 3.

While you're thinking about that second picture in Set IV, notice how slight a change necessitates an added path for traversing of the network. Figure 13 shows two such small alterations in that network, the first of which shows a tear at a vertex, convincing proof that the number of arcs is not involved in traversability. Examine both those changed networks for something about the degree of the vertices compared with that of the original in

FIGURE 12

Networks of One Vertex

Networks of Two Vertices

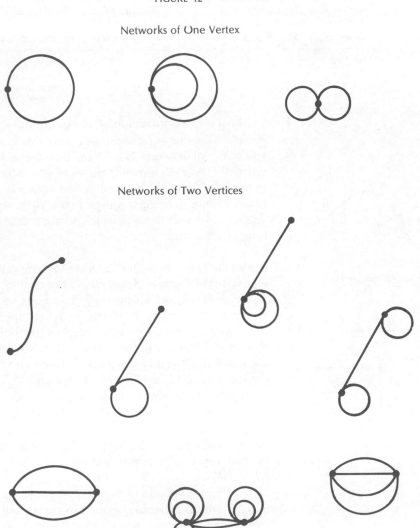

Set IV; this will give you a chance to check your sentences about those networks that can be traversed in one path.

Check the list of your sentences in Exercise 7b, completing to where you have at least five of them, no matter how trivial and obvious you thought they were. The next five phrases describing connected networks each fit in the blank in the sentence, "All networks that . . . can be traversed in one path.": "are simple closed curves," "are simple open curves," "have two or less vertices," "have all vertices of degree 2," and "have all even vertices." Each of those you found by yourself you probably understand well enough to convince another—to prove it to him by giving logical reasons why it must be so. A statement that can be proven within the mathematical system is called a _theorem_. Each of the five sentences just referred to is a theorem. (Granted that they do overlap and that they are not all marvelously important, they are still theorems.) Hopefully you had already discovered most of the five theorems, and you had mentally proved each to yourself even before you wrote it down on a permanent list.

If there is any theorem not yet proven for you, stop right now and reason why it must be so.

If you discovered by yourself the theorem that connected networks having all even vertices can be traversed in one path, then you also noticed something more about the nature of this single path. Completing the sentence for a more powerful theo-

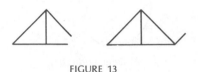

FIGURE 13

One-Room Cabin

FIGURE 14

rem, you should have something in your collection similar to Theorem 1.

THEOREM 1

If a connected network contains only even vertices, it can be traversed in one path starting from any vertex; and the path will be closed.

You may have several more theorems in your list for Exercise 7b, including one dealing with big networks having some odd vertices. Examining your statements so far, you may find that you have enough knowledge to be able to state the grand summarizing theorem about all sizes of networks; this is a completion of the following sentence: "If a network is traversed by a single path, then. . . ." Return to use your list in Exercise 7b plus your new ideas, and enter there any and all sentences that begin as above, "If a network is traversed by a single path, then. . . ." Let us test any rule you stated in Exercise 8 by examining some aspects of the networks for the one-room cabin and the seven bridges of Konigsberg. (Figure 14)

Observation of the network for the one-room cabin confirms the fact that we can traverse the network by starting either outside or inside of it (at 0 and _I_). If you choose to start at 0, you can tell by noting the degree of 0 that you will have to return once to 0. In attempting the network of Konigsberg bridges in one continuous walk, note that the town's walkers who start in the east part of town (E) will have to return once to E and leave again. This you know because the degree of E is 3. So, one who starts a walk at E cannot end the walk at E. Can he end his walk at a vertex of degree 2? Or 4? No, because an even vertex indicates that he must leave each time he enters that vertex. Where can he end his walk if he starts at E? He must end his walk at another odd vertex. There are too many odd vertices in that figure; he cannot traverse Konigsberg in one path.

Placing these observations in a more general context, consider the properties of a network that can be traversed in one path. This path must pass through all vertices between the first and last, and these middle vertices must be of a kind that you could only pass through—not end at. So, with the possible exception of the initial and terminal vertices, all the vertices would have to be even vertices. Looking again at the network for Konigsberg bridges (Figure 14), we see easily that there are four odd vertices. As Euler did, we conclude that it is impossible to promenade throughout the town in one path.

Notice that no matter how many vertices the network has, if it can be traversed by a single path, then it has no more than two odd vertices. This statement is a theorem—so called because it can be proven mathematically. Check this theorem against your most general sentence in Exercise 8. If you arrived at this theorem through your own rea-

Figure 14 Continued

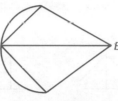

Ancient Konigsberg

soning, you can surely convince a doubting friend that it is true for any network he puts before you. If it is a bit new to you, check yourself by making a logical argument supporting the sufficiency of one path for traversing Figure 10c with its two odd vertices. But do note that your argument for a particular network is not a mathematical proof, for a mathematical proof must convince the reader without reference to any one network. You would not be expected at this stage in your mathematical maturity to write up a smooth general proof, but you will find it interesting to read this one.

Following is the grand theorem and a proof worded by B. H. Arnold on page 34 of *Intuitive Concepts in Elementary Topology*.[7]

Theorem 2

If a network can be traversed by a single path, then, with the possible exception of two of the vertices, each vertex of the network is even.

Proof

Let $a_1, a_2, \ldots, a_n$ be a sequence of arcs forming a path which traverses a given network, and let A be any vertex of this network except the initial and terminal vertices of this path. The initial and terminal vertices of the path may coincide or they may by distinct. We shall show that A is an even vertex of the network. Imagine

[7]B. H. Arnold, *Intuitive Concepts in Elementary Topology* (Englewood Cliffs, N.J.: Prentice-Hall, Inc., 1962), reprinted by permission of Prentice-Hall, Inc.

a point which starts at the initial vertex of a_1 and moves along a_1 to its terminal vertex (which is also the initial vertex of a_2), and then moves along a_2 to its terminal vertex (which is also the initial vertex of a_3), etc. until it finally arrives at the terminal vertex of a_n. Each time this point passes through the vertex A, it accounts for two arc ends at A—one on which to arrive and one on which to leave. Thus, the total number of arc ends at A must be even, and A is an even vertex of the network.

If you were forming theorems all the time, you would also realize that networks of no odd vertices can be traversed in a single path. There are several descriptions of networks guaranteeing that such networks can be traversed in one path, yet Theorem 2 has said nothing about this. If Konigsberg builds more bridges so that we have two odd vertices or no odd vertices, is there nothing we can be assured of? (As a matter of fact, Konigsberg has built two more bridges. See Exercise 9). Theorem 1 answers the question for networks having no odd vertices. Examine anew networks of two odd vertices such as those in Set IV and in the network of the Manhattan–New Jersey bridges (Figure 5). Notice that you are able to traverse all of them in one path as Theorem 3 states, and that, in order to do so, you pick an odd vertex as initial vertex. Theorem 3 and its proof are quoted directly from Arnold.[8] If you think that there is a parallel theorem for networks of one odd vertex, state and prove it yourself.

---

[8]Ibid., p. 35.

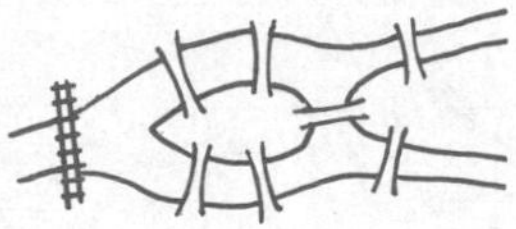

FIGURE 15

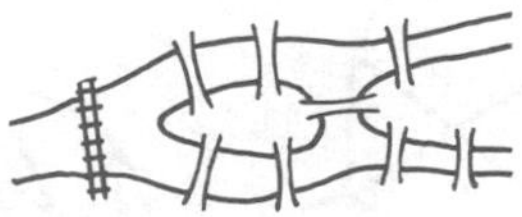

FIGURE 16

Theorem 3

If a connected network has exactly two odd vertices it can be traversed by a single path whose initial and terminal vertices are the two odd vertices of the network.

Proof

Given a network in which A and B are the only odd vertices, form a new enlarged network by joining A to B with a new arc a_0. In this enlarged network, every vertex is even; by Theorem 1.3, there is a path $a_0, a_1, \ldots, a_n$ that traverses this enlarged network. Then the path $a_1, a_2, \ldots, a_n$ traverses the original network and the initial and terminal vertices of this path are the two odd vertices A and B.

EXERCISE 9:

Later in history, Konigsberg added some bridges and changed its name to Kaliningrad.

a)   Can a clever resident make a walking tour of the city also using the new railroad bridge across the main Pregel? (Fig. 15) (Use the theorem to abbreviate your answer by referring to the number of odd vertices in the network as support for your answer).

b)   There are nowadays nine bridges in the city as shown in Figure 16.

Is it possible to take a walking tour now? (Again use the theorem in answering.)

Can it be done if the railroad bridge is excluded?

EXERCISE 10:

A network is given as the answer to Exercise 2b. Use a theorem in explaining why the network can or cannot be traversed in a single path.

How many paths are needed to traverse the network?

Can you form a rule based on the number of odd vertices?

EXERCISE 11:

For each of the networks in Figure 17 mark and count the odd vertices and prophesy how many paths will be needed to traverse the network. Check your prophecy by tracing out the paths.

Have you formed a rule for the number of paths necessary to traverse a network of x odd vertices?

How many are necessary for 6 odd vertices?

For 8 odd vertices?

For $2n$ odd vertices?

Can a network have an odd number of odd vertices?

From observations in Exercises 10 and 11, you have probably formed the following theorem.

Theorem 4

If a connected network has exactly $2n$ odd vertices, it can be traversed by a collection of n paths and cannot be traversed by any fewer than n paths.

◑ Note to Prospective Teacher:

A polygon is a particular network—not a region. Teachers can prevent their pupils from confusing a polygon with its enclosed region. Careful phrasing of verbal directions is a big help. A square cannot be colored blue—not really; yet one can trace around a square with a blue crayon and one can color inside a square. This is because a square is the union of four line segments, and line segments have no width. It is important to avoid any ambiguity in the mathematical use of a term. Asking students, "What is the perimeter of this square?" is free of ambiguity, for squares do have perimeter. But only after the students indicate a desire to establish a short convention should such idioms as "area of a square" be used; until then, the teacher should request "area of the region within the square," "area of this square patch," etc.

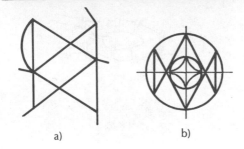

a)

b)

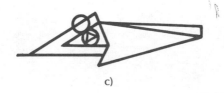

c)

FIGURE 17

Notice in Figure 18 the display of networks of degree 2. Networks b and d differ, in a way, from the others—differ in a way that is nontopological but will become important as our study becomes more specific.

The networks in Figure 18b and 18d are exceptional in that their arcs are all line segments. They are polygons.

A *polygon* is a network whose vertices are all of degree 2 and whose arcs are all line segments. Polygons are further classified for quick reference according to the number of line segments—as a triangle, a quadrilateral, a pentagon, a hexagon, a septagon, an octagon, etc. The polygon need not lie in a plane. A quadrilateral, for instance, can be formed from any four points in space. ◑

EXERCISE 12:

Is it possible to have a polygon lying on a plane?

On a sphere?

On a cylinder?

On the surface of a tetrahedron (a pyramid built on a triangular base)?

Polygon can also be defined as a simple closed curve made up of a union of more than two line segments. Both definitions emphasize that the polygon is a special type of curve, rather than a kind of surface. Both definitions also build upon topological concepts and point toward consideration of the number and kind of arcs. All polygons are topologically the same, but their various forms are of interest when later we explore more specific systems such as Euclidean geometry.

Polygons are not the only type of network that warrants special consideration because of particular relevance to other aspects of geometry. Examining the *skeleton* of a tetrahedron, we can see that the edges form a network. (In fact, all the skeletons you constructed with pipe cleaners in the first few chapters of the text were physical models of networks.) By an elastic motion, you can embed the tetrahedral network in the plane. The tetrahedral network is topologically equivalent to a network in the plane— that network seen in the third picture of Set IV of Figure 9. A network that is topologically equivalent to some network in a plane is called a *planar network*[9]. The tetrahedral network and all polygons are planar networks.

A printed circuit board printed on only one side is, of necessity, a planar network. The connecting wires are printed in a nonconducting material, and since the printed wires are not insulated, no two of them must cross except where a junction is intended. The wires are arcs of a planar network in

which care is taken that arcs meet at a vertex and do not approach each other when a junction is not intended. (Incidentally, the majority of circuit boards are printed on both sides, being models of little planar networks connected by wires through holes in the board.)

The following exercise and two others (Exercises 23 and 24 at the end of the chapter) will give you increased understanding of planar networks.

EXERCISE 13:

Do a cube's edges form a planar network? If so, be prepared to show the elastic motion. Experiment with pipe cleaners or, better yet, with two rubber bands held by four fingers of each hand—for you must visualize eight vertices of degree 3.

FIGURE 18

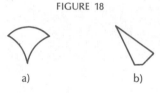

a)   b)

c)   d)

e)

5.4 EXERCISES FOR CHAPTER 5

14. a) Explain how you decided which points were vertices in Figure 10c. Is the decision you made the only possible one?

b) In a figure in which there is a point that may be a vertex but does not have to be a vertex, what is the degree of this point if it is considered to be a vertex?

15. For each of the following, if it is possible to do so, draw a network that can be traversed in one path and has the given number of vertices

---

[9]We shall call a network "planar" if it can be embedded in the plane; some authors will use the term to mean that it is already embedded in the plane.

and the given number of arcs.

a) 3 even vertices and 3 arcs

b) 3 even vertices and 6 arcs

c) 3 even vertices and 5 arcs

d) 3 even vertices and 3 odd and 6 arcs

e) 5 even and 2 odd vertices

f) 8 odd vertices

16. Explain why a network with exactly two vertices cannot have one even vertex and one odd vertex. (See Exercise 11).

17. An airline has routes serving four cities: A, B, C, and D. They lie on a map as in Figure 19a. The director wishes to present his pattern of routes diagrammatically and chooses to show his routes as straight lines in displaying the network (Figure 19b). He could have expressed himself more flamboyantly and drawn networks that seem different at first but that are not topologically different. (Figures 19c, d, e)

a) Try to find networks different from the director's in which B is of degree 3, C of degree 1, and the others of degree 2. (There is at least one.)

b) Can the director adequately describe his service by stating in words how many routes come into each city (how many arcs are at each vertex)? Try it!

18. Consider this chemical application from Arnold. The structure of a molecule can be

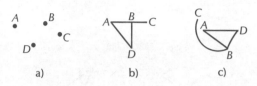

a)   b)   c)

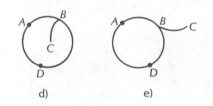

d)   e)

FIGURE 19

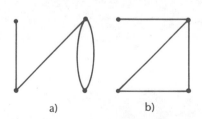

a)   b)

FIGURE 20

schematically represented by a network. The vertices of the network represent the atoms of which the molecule is composed and the arcs represent chemical bonds between certain pairs of these atoms. Figure 20 shows two essentially different ways of forming a molecule with four atoms, two of the same element having two chemical bonds each while the other two have one bond and three bonds, respectively.

a) Prove that the molecules diagrammed in Figure 20 are the only ones that can be formed from four atoms with the given bonds.

b) Find three molecules that can be formed from four atoms, two of which have two chemical bonds each, while the other two have three bonds each.

c) Find four different molecules that can be formed from four atoms, two of which have three chemical bonds each, while the other two have two and four bonds, respectively.

19. Determine whether you can come up with several networks of four vertices, each of which is of degree 3. Remember from Exercise 14 to ignore vertices of degree 2, since corners can always be rounded.

20. Is it true that two networks are equivalent that have the same degree at the respective vertices? If false, you can find a counterexample— an instance where the specifications are fulfilled and the two networks are not equivalent.

If true, careful reasoning can substantiate its truth. Prove it true or false.

21. Relative to exploring the answer to Exercise 20, Alan Tammadge[10] suggests in an article that a table can be drawn up showing the number of different direct routes from the vertices in the column at the left to the vertices in the row at the top. As an example, Figure 21a contains a network having 4 vertices of degree 3, and Figure 21b shows the table representing the routes of that network. He suggests that the generation of tables will help a student find new networks (if possible) and perceive equivalent networks.

a) Do the tables always yield a network? The table in Figure 22 does not yield a network. Why?

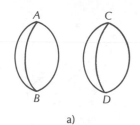

a)

	A	B	C	D
A	0	3	0	0
B	3	0	0	0
C	0	0	0	3
D	0	0	3	0

b)

FIGURE 21

FIGURE 22

	A	B	C	D
A	3	0	0	0
B	2	0	0	0
C	0	0	3	0
D	0	0	0	3

b) Try forming a network of a vertex of degree 4, a vertex of degree 3, and a vertex of degree 5. You'll find that it can be done. A clever way is to leave a hole in the middle of an incomplete sketch as in Figure 23, and then

[10] Alan Tammadge, "Networks," *The Mathematics Teacher* (November 1966), p. 624–29.

FIGURE 23

FIGURE 24

reason and experiment as to the connecting of the end points. By playing around with this sketch, you will find that so slight a change in specifications as to make the last vertex a vertex of degree 4 or a vertex of degree 2 renders the drawing of the network impossible; you may find one arc with nowhere to go but out in space, necessitating the addition of a vertex of order 1. (The reason concerns the total arc degree of the networks discussed in part of Exercise 11.) Read Tammadge's article for more on this approach to make up networks to certain specifications.

Find and state a rule as to the type of vertices composing a realizable network.

22. a) What relation holds between the number of vertices and the number of sides of a polygon?

b) Figure 24 shows a quadrilateral in which a dotted line has been added to break the quadrilateral into triangles. Can any quadrilateral region be shown as a union of adjoining triangular regions? Can any pentagon? Any hexagon? Any polygon—even a weirdly-shaped one?

c) The polygonal surface in Figure 24 can be expressed as a union of two triangular surfaces. In those polygonal surfaces expressible as a union of triangular surfaces, is there any rule as to how many triangular surfaces are bounded by a polygon of n sides?

23. Notice that a triangle can never be drawn so that it does not lie in a plane. Hold a polygon of three sides in space, and you will see that you can *always* find a plane containing that particular network.

Is this true for a polygon of four sides?

Of five sides?

24. Nonplanar networks occur in practical situations. One example of a nonplanar network is the network of supply lines for gas, water, and electricity (G, W, and E) to each of three homes (labeled A, B, and C). A network is <u>nonplanar</u> if it is not equivalent to a network in the plane. That is, in the squashing and stretching of a nonplanar network to force it into the plane, a nonequivalent network of additional vertices would be formed.

a) Finish the connections started in Figure 25.

Try connecting each house to each utility main, G, W and E so that no utility line passes through two houses or crosses another line.

b) Why is it impossible to make all the connections in the plane?

c) Can the gas-water-electricity network be drawn on the surface of a sphere? Of a torus?

5.5 APPLICATIONS OF NETWORKS

Finding one's way from one place to another, as along a network, is the basis of many puzzles. We

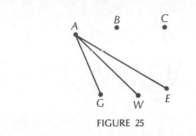

FIGURE 25

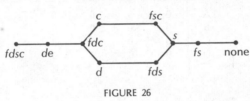

FIGURE 26

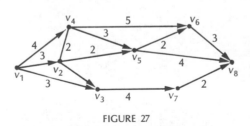

FIGURE 27

can illustrate still another use of networks for finding or enumerating possibilities. Let us use the ancient Ferryman's Puzzle to illustrate what we have in mind. A ferryman f has been given the task of bringing across a river a dog d, a sheep s, and a bag of cabbage c. His little rowboat can carry only one item at a time, and obviously he cannot leave the dog alone with the sheep, nor the sheep with the cabbage. How shall he proceed?

We analyze the various alternatives and agree that he has to bring the sheep over first. This leaves d and c on the bank where originally f, s, d, and c all stood. As you continue, you realize that there are alternatives when he returns to stand with d and c on the bank. To keep track of the action, you could benefit by drawing a network whose vertices represent each change in those standing on the first bank. (See Figure 26).

In management, the earliest a task (in a succession of tasks) can be scheduled and the latest it needs to be scheduled can be determined by making and studying a network of the total project in which each specific task is one vertex. In the "flow" network in Figure 27, v_1 represents the beginning of task 1 and v_8 the beginning of the task that cannot be started until all the others have been done. Optimum scheduling of v_8 is associated with the longest path, and the scheduling of v_3 and v_7 can be given a little slack since they are not on this longest ("critical") path.

Let us attach numbers of hours to the arcs to indicate the duration of activities and read a few facts

⊕ Note to Prospective Teacher:

The following is an application of networks to first efforts at map drawing.

Children's maps are generally topological equivalents to city maps. Yet children's maps are seldom congruent to maps that city fathers authorize for publication. Children often see their routes in terms of right and left turns and landmarks; distance and the sharpness of a turn are not important. (See Figure 28 for sketches of two routes home from school.)

The Nuffield Mathematics Project in Great Britain suggests that geography and local studies lead into network maps, which are the child's first efforts at communicating his experience in coming to school. In the booklet, *Shape and Size 2*, the people of Nuffield Project give several suggestions for these first stages of drawing maps.[11] A start can be made by having the children observe things on the way to school such as houses, shops, roads, road junctions, factories, and bridges. A visual record can be made that will at first be no more than the children's own imaginative interpretation of their way to school. But this can open some interesting discussion on the part of the children about the way they come to school so that they get practice in giving verbal directions—without having to consider any idea of distance or scale. (Adults, too, rarely give indication of distance; rather we use such expressions as "Take the first right and then turn left at the Esso station.") With children, discussion can arise from questions such as, "How many roads do you have to cross on the way to school? Could you keep your parents less worried about you by planning another

route that does not involve crossing big streets?" and, "What other way could you go home if that street were closed for repairs?" Definite points on the way to school can be located and routes planned to include for example, "Which way would you go home if you wanted to buy a candy bar and play on the big slides and swings?" (The answer to this is the longest route of the network in Figure 28). Although some children will first have to watch the teacher making network maps, Nuffield claims that each will be able to attempt these after discussion about alternate routes and attempts to give verbal directions to a friend.

These are only a few suggestions; the teacher will think of many other suitable to his environment. Note that the drawings are properly networks because at this stage no mention is made of measurement or distance.

FIGURE 28

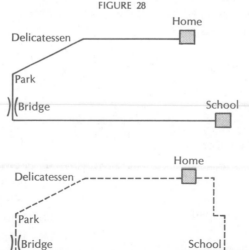

Usual Way Home

from Figure 27. The final task, v_8, cannot be started until 13 hours have elapsed, v_4 cannot begin until 5 hours have elapsed, and v_7 can commence as early as the eighth hour and as late as the eleventh hour. If you so desire, you can give yourself exercises in interpreting and constructing critical path networks.

Optimizing routes of travel and speed of transmission of electrical current usually involves minimizing the number of interchanges. Schematizing familial and hierarchical relationships, showing possibilities of verbal communication between people in a small group, and diagramming grammatical functions of words in sentences—these are all familiar uses of networks. There are a host of other applications including game analysis and models of strategy in war and disarmament. For interesting reading on applications of networks, open the text by Busacker and Saaty to Chapter 6. (See bibliography) ⊕

[11] Nuffield Mathematics Project, *Shape and Size 2* (New York: John Wiley & Sons, Inc., 1968), pp. 79–80.

ANSWERS TO EXERCISES

1. Networks: b) with 9 arcs and 6 vertices.
   c) with 8 arcs and 5 vertices.
   d) with 12 arcs and 7 vertices.
   e) with 3 arcs and 4 vertices, 2 of which are of degree 2.

2. a) I ⊂⊃ O

   b)

   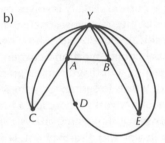

   c) The network you draw should have four vertices and seven arcs. The question of the walk is progressively easier to answer as you continue studying.

3. a) Yes, if you can draw it like a fantastic daisy: the following shows eight arcs and one vertex.

   b) No.   c) Here is one: Y//

4. a) All but two of the networks can be transversed in one path; two networks in set IV require two paths.

   b) Not all open curves; not Y. All simple open curves can be traced in one path.

   c) All simple closed curves can be traversed in one path, but not all closed curves.

5. Four paths needed to traverse Copenhagen's train system.
   In Figure 4, b can be traversed in two paths, c in one path, d in two paths, and e in three paths.

6. For two figures,

Figure 10b	Figure 10c
a) (ADFECBC) (DE)	(CABCJEFGKEDJHKF)
b) (1323323) (31)	(322344324424243)

8. a) No.

9. Yes; two vertices of degree 4 and two odd vertices.
   b) Yes. Yes.

10. Two paths needed because the network has more than two odd vertices.

11. In Figure 17a, 5 vertices of degree 1
    4 vertices of degree 3
    2 vertices of degree 4
    1 vertex of degree 5
    1 vertex of degree 6
    10 odd vertices, so 5 paths needed.

    In Figure 17b, 14 vertices of degree 4, so one path suffices.

    In Figure 17c, 1 vertex of degree 1
    4 vertices of degree 3
    5 vertices of degree 4
    1 vertex of degree 5
    1 vertex of degree 8.
    6 odd vertices, so 3 paths needed.

    General: 3 paths for 6 vertices; 4 paths for 8 vertices; n paths for $2n$ odd vertices. We can't have 9 odd vertices because every arc has 2 vertices, so the total degree of all the vertices must be even.

12. Polygon can't lie on the surface of a sphere or cylinder because the line segments won't lie in the surface. Can lie on a plane and on some places on a tetrahedron.

13. Cube's edges are a planar network.

14. Vertices of degree 2 can be omitted or included, as you choose.

15. a)

    b)

    c)

    d) impossible to form such a network.

    e)

    f) impossible to traverse such a network in one path.

16. See answer to Exercise 11.

17. a) Figure 20a happens to be such a network.

17. b) No, since two readers could get different conceptions. The number of arcs at a vertex does not determine the network, as you just demonstrated.

19. There are eight of them. Figure 21a shows one of them.

20. Proven false by any counterexample; a nonequivalent pair can easily be found in working on Exercise 19.

22. The number of sides equals the number of vertices, because they are all of degree 2.

23. A polygon can have its vertices any place in space; hence one vertex of a quadrilateral could be outside of the plane containing the other three. Similarly for five-sided polygons.

BIBLIOGRAPHY

Arnold, B. H. *Intuitive Concepts in Elementary Topology.* Englewood Cliffs, N.J., 1962, pp. 21–43.

Ball, W. W. R. *Mathematical Recreations and Essays.* Revised by H. S. M. Coxetor. New York: The Macmillan Co., 1960, pp. 242–66.

Busacker, Robert G., and Saaty, Thomas L. *Finite Graphs and Networks: An Introduction with Applications.* New York: McGraw-Hill Book Co., 1965.

Cantor, Moritz. *Vorlesungen uber Geschichte der Mathematik.* Vol. 3. Leipzig, 1901, p. 552.

Cairns, Stuart Scott. *Introductory Topology.* New York: Ronald Press Co., 1961.

Cornwell, Bruce and Katherine. "The Seven Bridges of Konigsberg." International Film Bureau, 1965.

Denholm, Richard. *Mathematics: Man's Key to Progress.* Pasadena, Calif.: Franklin Publications, Inc., 1968, p. 74.

Euler, L. *Opera Omnia.* Edited by Ferdinand Rudio, Adolf Krazer, and Paul Stackel, Leipzig: B. G. Teubneri, 1911–62.

Fujii, John N. "Puzzles and Graphs. "*National Council of Teachers of Mathematics.* Washington, D.C., 1966.

Haray, Frank. *Graph Theory.* Reading, Mass: Addison-Wesley Publishing Co., 1969.

Hartung, Maurice L. and Stochl, James E. *Foundations of Elementary School Arithmetic.* Glenview, Ill.: Scott-Foresman & Co., 1965, pp. 379–91.

Kasner, Edward, and Newman, James. "Rubber-Sheet Geometry." *Mathematics and the Imagination.* New York: Simon & Schuster, Inc., 1963, pp. 265–77.

Loyd, Samuel. *Best Mathematical Puzzles of Sam Loyd.* Edited by Martin Gardner. New York: Dover Publications, Inc., 1959.

Meyer, Walter, "Garbage Collection, Sunday Strolls, and Soldering Problems." *The Mathematics Teacher* (April 1972), pp. 307–9.

Newman, James R., ed. *The World of Mathematics.* Vol. 1. New York: Simon & Schuster, Inc., 1956, pp. 573–80.

Nuffield Mathematics Project. *Shape and Size 2.* New York: John Wiley & Sons, Inc., 1968, pp. 79–80.

Ore, Oystein. *Graphs and Their Uses.* New Mathematical Library. New York: Random House Inc., 1963.

Ptak, Diane M. *Geometric Excursion.* Oakland County Mathematics Project. Oakland, Mich. 1970. p. 21.

Roper, Susan. *Paper and Pencil Geometry.* Newport Beach, Calif.: Franklin Publications, Inc., 1966, Unit 12.

Rosskoph, Myron F.; Levine, Joan L.; and Vogeli, Bruce R. *Geometry: A Perspective View.* New York: McGraw-Hill, Inc., 1969, pp. 259–60.

Speiser, Andreas. *Klassische Stucke der Mathematik.* Zurich: Orell Fussli, 1925, pp. 127–38.

Struik, Dirk Jan ed. *A Source Book in Mathematics: Twelve Hundred to Eighteen Hundred.* Cambridge: Harvard University Press, 1969, pp. 184–7.

Tammadge, Alan. "Networks." *The Mathematics Teacher* (November 1966), pp. 624–9.

Turnbull, Herbert W. "The Bernoullis and Euler." *The World of Mathematics.* Edited by James R. Newman. New York: Simon & Schuster, Inc., 1956, pp. 147–51.

Van Engen, Henry; Hartung, Maurice, L.; and Stochl, James E. *Foundations of Elementary School Arithmetic.* Glenview, Ill.: Scott-Foresman & Co., 1965, pp. 379–393.

chapter 6

maps

MATERIALS NEEDED

Models of a tetrahedron and a cube

A large ball capable of showing chalk marks

Two pieces of cloth big enough to drape around that ball

Optional:

Crayons and colored chalk

Globe

A *map* is a surface in which lies a network. In this chapter you will be using maps on a plane, first of all, to determine optimum map coloring, and secondly to deduce the relationship between the number of faces, edges, and vertices of any connected planar map elaborated in Euler's formula. There is also a section in this chapter devoted to the study of polyhedra, defined here as a special map.

6.1 MAP COLORING

How many colors do you really need to color a map drawn on a sheet of paper? No one knows for sure! Given certain ground rules in coloring such a map, it appears that there is a least number of colors needed, but no one can prove that this number will suffice.

Pretend that Figure 1 is a map of 4 countries, one

FIGURE 1

a)

b)

c)

d)

e)

f)

FIGURE 2

in the center and three banding it. The rules of map coloring are that no 2 countries sharing more of a border than a point can be colored the same color. So the figure certainly seems to need 4 colors for the 4 countries.

By convention, we color the map as if it were alone in the plane and as if the surface outside of the plane—like the sea—were assigned a color. Following this convention, we see that the sea can be assigned the same color as the center country, and hence conclude that 4 colors still suffice.

EXERCISE 1:

Try your hand at map coloring so that no two regions sharing a border are the same color. For each map in Figure 2, indicate how you would color the map using the fewest colors.

You could color some of the maps with 2 colors, or 3, but not all of them. As you noted in Figure 2, it is pretty easy to demonstrate maps that require 4 colors. Figure 3 shows a region of Europe needing 4 colors just to color Luxembourg and its bordering countries. Would a fifth color be necessary if we finished the map by assigning a color to the region of water lying west of the coastline?

A *map* is composed of a network and a surface containing the network. The network's *arcs* and *vertices* become those of the map, and the map has faces that are the regions isolated by the network. We will start out by considering only *planar* _maps_, maps drawn in a plane. A _planar map_ is the surface associated with a planar network. You recall perhaps that a planar network is one that can be drawn in the plane in such a way that its arcs intersect only at their endpoints.

The distinction between a network and a map is one of viewpoint, as is perceiving background instead of foreground. In a map, interest is focused on the portions into which a surface is divided by the arcs of the related network, with the network itself playing a subordinate role. In an ordinary map in an atlas, these portions of the surface are the states or countries; in the general case, these portions of the surface are called _faces_. By convention, the portion of the surface "outside" the network is counted as one face of the map; thus, in a plane, one of the faces of a map will be unbounded. The arcs and vertices of the network are called _edges_ and _vertices_ of the map, respectively. Note that, although edges usually lie between two faces, it is possible to have an edge with only one face. (The vertical line segment connecting parts of Figure 2e is a single edge in the unbounded face.)

In coloring a map, two faces having an edge in common must be colored with different colors; if only a point in common, they can be colored the same color. We will abide by the convention of assigning color to the region outside. Hence, in Figure 2a only two colors are needed for the four sectors of the circular region since the same color can be used for the diagonally opposite sectors. The region exterior to the circle would have to be colored a third color.

Figure 2 Continued

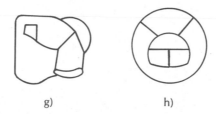

g)   h)

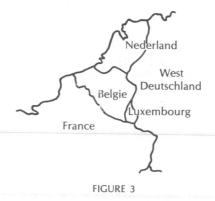

FIGURE 3

EXERCISE 2:

Copy the map in Figure 3 and color _all_ faces including the "region outside," keeping to a minimum the number of colors used. Also, check and improve your coloring of Figure 2. Do you need 5 colors for any map in Figure 2?

It is provable that every map in a plane can be colored with 5 colors, but no one has yet found a map for which 5 colors are needed—in each case so far, a map can be colored with 4 colors. Try dreaming up your own maps or making variations on those given in Figure 2; you will learn a great deal and you may see a way of arranging the pattern so that 5 colors are needed. If so, your name will go down in history. There is only one hitch: it has been proven that 4 colors are sufficient for planar maps of less than 40 faces.

Several excellent topologists have worked on the idea of proving that 4 colors are sufficient for all planar maps, but no proof has yet been found. It is especially irritating that the sufficiency of 4 colors for all planar maps cannot be proven when in fact there are comparable theorems applicable to complicated surfaces such as the Mobius strip and the torus. The sufficiency of 6 colors for a Mobius strip and 7 colors for a torus has been proven. (See Barr[1] for pictures of these colored surfaces.) Although we cannot yet prove the sufficiency of 4

[1]Stephen Barr, _Experiments in Topology_ (New York: Thomas Y. Crowell Co., 1964).

colors for a planar map, so far it has always been possible to color one with 4 colors, much to the frustration of those who would like an example showing five colors to be necessary. Each time that a fifth color seems necessary, you can backtrack and reassign colors to result in a finished map of only 4 colors.

EXERCISE 3:

Here's a two-handed game that will give you insight into map coloring. Player A draws a region. Player B colors (or labels) it and draws a new region. Player A colors it and adds a third. This goes on until somebody gets stuck and has to use a fifth color due to mismanagement. Traps can be set for the naive—and sometimes avoided. Play it at home for 20 minutes, before challenging any of your sophisticated classmates. (For a more rewarding game, see Exercise 11).

6.2 EULER'S FORMULA

There is a most fascinating relationship between the number of faces, edges and vertices of any connected planar map. (There are also set relationships for maps on a torus, on a Klein bottle, and on a Mobius strip—that is, for each surface, there is a certain numerical relationship between the faces, edges, and vertices).

EXERCISE 4:

a) Survey the maps in Figures 1, 2, and 3 with an eye to discovering the simple numerical relationship between faces, edges, and vertices in a planar map. To do this, tabulate the number of faces, edges, and vertices for each of the ten maps involved; examine your numbers for a consistent comparison between the number of edges and the sum of faces and vertices, thus completing the sentence, "Number of faces plus number of vertices equals...."

Name or sketch of map	No. of		
	Faces	Vertices	Edges
Figure 1			
Figure 2a			
Figure 2b			
Figure 2c			
Figure 2d			
Figure 2e			
Figure 2f			
Figure 2g			
Figure 2h			
Figure 3			

$$n(F) + n(V) = ?$$

b) If your formula is correct, it should work for any connected planar map. A map is said to be _connected_ if, and only if, the network of the map is connected. Check your proposed formula by sketching two connected maps that are radically different from the maps in Figure 2, writing below each, the number of faces, edges, and vertices, respectively.

FIGURE 4

In Figure 4, there is a map of just one face, one edge, and two vertices: $F + V = 1 + 2 = E + 2$. In general, 2 is the magic total for the number of faces plus the vertices minus edges.

FIGURE 5

THEOREM (EULER'S)

If F, V, and E are, respectively, the numbers of faces, vertices, and edges, of a connected planar map, then $F + V = E + 2$.

DISCUSSION

Given a connected planar map, there is a certain number of faces plus vertices minus edges. It will be proven here that this number can be maintained through certain simplifications of the map down to a single edge and 2 vertices in 1 face, where we can see (Figure 4) that $F + V = E + 2$, or $F + V - E = 2$.

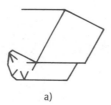

a)

PROOF

Consider a connected planar map, M, as pictured in Figure 5a. For this map, let $F + V - E = N$, where N is some number. A vertex can be removed, from the map at the point where only two edges meet at one vertex; this results in a

b)

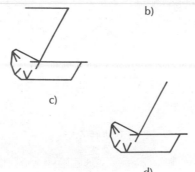

c)

d)

change analogous to that pictured between Figures 5a and 5b, in which the loss of one vertex and one edge leaves $F + V - E = N$. Removal of an edge separating two faces, as in the change to Figure 5c, results in no change in the formula since an edge and a face have both been removed. Deletion of an edge with one free end makes a change in the map like that seen in simplification from Figure 5c to 5d. The number, $F + V - E$, remains the same because both an edge and a vertex have been subtracted.

Successive removals of vertices and edges as specified above can be continued throughout, so long as the connectedness of the map is maintained. The final simplification yields the single edge and single face seen in Figure 4. Since N has not changed in any of these successive simplifications, $F + V - E = 2$ in the original map as well. $F + V = E + 2$ for any connected planar map.

Connected planar maps are said to have an _Euler number of 2_. The Euler number of a surface is a topological property of that surface. Euler's idea can be generalized to include any planar drawing that is in lines and dots: one dot on a sheet of paper, $F + V = 1 + 1 = 0 + 2$, and disconnected maps having n separate parts, $F + V - n + 1 = E + 2$.

Euler's theorem relating to planar maps is used to prove that any planar map can be colored with 5 colors, for the theorem allows a topologist to arrive at a limitation of the complexity of one of the faces—

that in a planar map there is at least 1 face with 5 or fewer edges. (For this proof, see footnoted material on the 4-color problem.)[2]

EXERCISE 5:

Count the number of faces, edges, and vertices for a "squared doughnut" as shown in Figure 6. In cases of doubt, a face is defined as a region topologically equivalent to the interior of a plane polygon. Does the "squared doughnut" have the same Euler number as the planar map? If you think of this is a map, on what surface is it drawn?

EXERCISE 6:

Report your count of faces, edges, and vertices of a cube and of a pyramid built upon a square base. What is the Euler number for each?

6.3 POLYHEDRA

Notice that the cube and pyramid are topologically spheres and yet have the same Euler number as planar maps. The Euler number of a map on a sphere

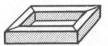

FIGURE 6

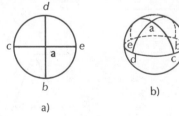

a)

b)

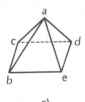

c)

FIGURE 7

is the same as that of a planar map. This is because the two maps have the same number of faces, vertices, and edges. A planar map can be draped around a sphere and then sewn together in some face without changing its coloring or Euler number. To see this, look at the planar map in Figure 7a. It has four inside faces and one outside face. Mentally drape it around a baseball and sew the seam in that face previously called the "outside face." (Figure 7b) In Figure 7b imagine the stitching in a pucker on the bottom, then do a little mental patting of the faces of the sphere to get the pyramid in Figure 7c. Hence, the colors you choose in coloring the square-based pyramid in Figure 7c will work on the sphere and on the planar map lying to its left. Identical map-coloring problems can be posed in the plane, on a sphere, and on the surfaces of certain familiar three-dimensional objects such as a closed shoe box.

In the same sense that the polygon is a special network, the _polyhedron_ is a special map. A simple _polyhedron_[3] is topologically equivalent to a sphere and is a map with each face being a piece of a plane (i.e. being flat). Since a map is a surface, a polyhedron is merely a certain kind of surface. The pyramid pictured in Figure 7c is a polyhedron— "poly" means many and "hedron" means face— and it is topologically equivalent to the surface in Figure 7b. Cubes and tetrahedra are also amongst the many _polyhedra_ (that's the plural of polyhedron).

---

[2] W. W. R. Ball, _Mathematical Recreations and Essays_, rev. ed H. S. M. Coxeter (New York: The Macmillan Co., 1960), pp. 222–9; H. S. M. Coxeter; "The Four Color Map Problem, 1890–1940," _The Mathematics Teacher_ (April 1959), pp. 283–89; B. H. Arnold, _Intuitive Concepts in Elementary Topology_ (Englewood Cliffs, N.J.: Prentice-Hall, Inc., 1962, pp. 43–53.

---

[3] We shall confine our discussion to simple polyhedra, and thus omit the adjective "simple."

EXERCISE 7:

a) Can a polyhedron sink in water?

b) Is a polyhedron a special network? Is a polyhedron a map?

c) Is it possible to make a polyhedron out of cardboard?

d) Can there be a polyhedron of three faces? of four faces?

e) Are there more vertices plus faces than there are edges to a polyhedron?

f) Is a sphere a polyhedron? Is a cylinder with both its ends attached a polyhedron?

g) Is the surface of a two-by-four piece of lumber a polyhedron?

h) Is the total surface of a topless wooden box a polyhedron?

i) What kind of curve is the edge of a polyhedron?

EXERCISE 8:

Which of the figures in Figure 8 do you think will satisfy Euler's Formula? Try to answer before counting.

EXERCISE 9:

Color a tetrahedron and a cube according to the rules previously set up for planar maps. Study the instructions on coloring maps, then write a sentence or two giving instructions to an imaginary classmate on coloring polyhedra. What changes, if any, are

FIGURE 8

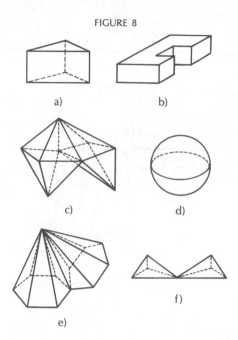

a)   b)

c)   d)

e)   f)

FIGURE 9

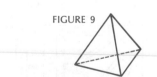

FIGURE 10

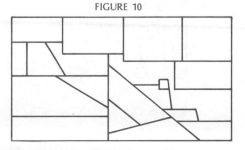

there in the wording of map-coloring instructions to apply to polyhedra? What changes in the concept?

EXERCISE 10:

Figure 9 portrays a tetrahedron.

a) Draw a sketch, topologically equivalent to it, on a sphere.

b) Draw a planar map equivalent to it.

EXERCISE 11:

Repeat Exercise 10 for a cube, sketching the cube yourself.

6.4 EXERCISES FOR CHAPTER 6

12. At home, prepare your attack for the following two-handed game. You will trade the partially colored map of Figure 10 with an opponent. The object is to, ahead of time, fill in regions in such a way that your opponent cannot manage to finish the map in 4 colors. The rules are that you use 4 or fewer colors and that you fill in as many regions as you wish of those regions that don't have a border in common. (Should you feel further motivated, copy the map and try coloring that copy with three or less colors so that you still trap an opponent.) In class, trade your colored map for that of an opponent and let the game proceed with a time limit.

13. Design a flashy modern stained glass window obeying map-coloring rules. Even the un-artistic can contribute something bold and colorful. Perhaps ideas will flow better if you first color maps of Figures 11a and 11b.

14. Color Figure 12, which is a map having all vertices of degree 3. This kind of map is called a *regular map*. Every map can be converted into a regular map by inserting a bubble at each vertex of degree greater than 3. (See Arnold for details; there the 5-color theorem is proven, the proof based on regular maps.) Try sketching and coloring regular maps converted from Figures 2a and 2b.

15. Examine classroom models of polyhedra, counting the faces and the vertices and projecting the number of edges using Euler's theorem. It is customary and easiest to count the largest polygonal surface in each plane as one face. If all the faces are exactly the same size and shape, we call the polyhedron *regular*. Point out the regular polyhedra. Count the faces, edges, and vertices of a certain regular *poly-hedron* one in which each face is a surface bounded by a regular pentagon. (A *regular pentagon* is one in which each line segment is of the same length.) How many faces, edges, and vertices has this regular polyhedron of regular pentagonal faces?

16. A concise way of explaining the relationship between a planar map and a map on a sphere

FIGURE 11

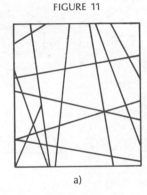

a)

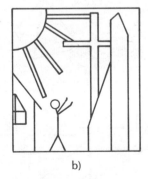

b)

utilizes the concept of projection from a point on the sphere (call it N). Visualize a *globe* sitting on a plane and call a point on the globe, the pole; let's use the North pole of a globe sitting on your desk as the point N. (see Figure 13.)

For any point P on the sphere, and different from N, the line through N and P intersects the plane in exactly one point Q. Conversely, for each point Q in the plane, the line through N and Q intersects the sphere in exactly one point P that is different from N. Thus we have a correspondence between all points in the plane and all but one point on the sphere.

The South pole, S, corresponds to itself in the plane. The point N on·the sphere does not correspond to any point in the plane, but N is the only point on the sphere with no corresponding point in the plane. This correspondence is called a *polar projection*.

Any map on a sphere can be transferred to a plane by choosing N to be any point not on an edge, and using the polar projection from N. Conversely, every map in a plane can be transferred to a sphere by a polar projection.

a) Draw a sphere with North Pole as N and sketch three or four curves on the sphere whose images in the plane are straight lines.

b) Do the same for two line segments sketched in the plane.

c) What are the images in the plane of latitude circles on the sphere?

d) Sketch a projection of a torus to the plane

or explain why such a transfer cannot be made.

17. Is the map of the "squared torus" also a planar map? (See the sketch of the "squared dough-nut" of Exercise 5).

18. Is the map formed from the *gas-water-electri-city network* a planar map? (See Exercise 24 in Chapter 5).

19. Is the map formed from the *complete network on five points* (Figure 14) a planar map? Explain, including an argument based on its Euler number. The *complete network of five points* is the network in which each of the five ver-tices is joined by an arc to each of the other four vertices. All the faces thus are triangles because each pair of vertices is connected. The number of combinations of five things taken two at a time is the number of edges.

FIGURE 12

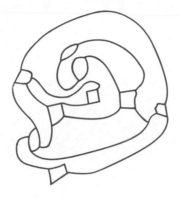

FIGURE 13

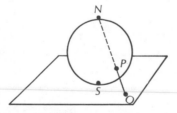

FIGURE 14

ANSWERS TO EXERCISES

1. Below are sample answers amongst many correct answers.

2. You should be able to color the map in four colors by making the sea the same color as Luxembourg. Likewise, 4 colors or less for each map of Figure 2.

4. a) In Figure 1, 5 faces, 6 vertices, 9 edges.
        In Figure 2d, 4 faces, 4 vertices, 6 edges.
        In Figure 2g, 6 faces, 8 vertices, 12 edges.

5. 16 faces: 4 outside and 4 inside, 4 on top and 4 on bottom; 32 edges and 16 vertices. We cannot count the whole top surface as one surface because it is not a plane polygonal region. Zero is the Euler number for this and all other maps on the surface of a torus.

6. Cube: 6 faces, 8 vertices, 12 edges.
    Pyramid: 5 faces, 5 vertices, 8 edges.
    Euler number is 2 on a spherical map.

7. a) No.   b) No; yes.   c) No; no thickness to a polyhedron.   d) No; Yes.   e) Yes, two more.
    f) No; No.   g) Yes.   h) Yes.
    i) A polyhedron's edges are all line segments.

9. Each face of a polyhedron must be colored, and two faces having an edge in common must be colored with different colors. If two faces have only a point in common, they can be colored the same color.

    Different only in that there is no fuss over a special face previously denoted as "the outside face."

    There is no change in the concept of map coloring.

10. b)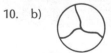

14. A regular map related to figure 2a.

15. It's not hard to cut out some regular pentagons from tracing around a model, and then tape five of them to the base and try to fold them up as if you were going to close them into a rough sphere. You'll see very quickly how many other pentagonal faces are needed, and thus how many vertices the polyhedron has. From there you can deduce the number of edges that would be present on the finished model.

16. b) arcs along a longitude curve on the sphere.

    c) circles in the plane.

17. No, for the Euler number is different for maps drawn on the torus.

18. No, a count of the Euler number will confirm this. The map has three faces as shown on this unrolled torus. Three faces, six vertices, and nine edges imply an Euler number of zero.

19. Nonplanar with an Euler number of 5.

    Ten faces: six triangles in the "almost planar" quadrilateral and four triangles from the point above center.

BIBLIOGRAPHY

Arnold, B. H. *Intuitive Concepts in Elementary Topology.* Englewood Cliffs, N.J.: Prentice Hall, Inc., 1962. pp. 43–53.

Ball, W. W. R. *Mathematical Recreations and Essays.* Revised by H. S. M. Coxetor. New York: The Macmillan Co., 1960, pp. 222–9.

Barr, Stephen. *Experiments in Topology.* New York: Thomas Y. Crowell Co., 1964.

Caldwell, J. H. *Topics in Recreational Mathematics.* New York: Cambridge University Press, 1966, pp. 76–87.

Courant, Richard, and Robbins, Herbert. *What is Mathematics?* New York: Oxford University Press, 1960, pp. 235–40, 246–48.

Coxetor, H. S. M. "The Four Color Map Problem, 1890–1940." *The Mathematics Teacher* (April 1959), pp. 283–9.

Euler, Leonard. Euler's Solution, a report to The Academy of Sciences of St. Petersburg. Vol. 8 of report of the Academy, 1736, p. 74.

Ore, Oystein. *Graphs and Their Uses.* New Mathematical Library. New York: Random House, Inc., 1963, pp. 109–16.

Polya, George. *Induction and Analogy in Mathematics. Mathematics and Plausible Reasoning.* Vol. 1. Princeton, N.J.: Princeton University Press, 1954, pp. 35–43.

Schaaf, William L. *Basic Concepts of Elementary Mathematics.* New York: John Wiley & Sons, Inc., 1966, pp. 241–3.

Tietze, Heinrich. *Famous Problems in Mathematics.* 2d ed. Baltimore: Graylock Press, 1965, pp. 226–42.

chapter 7 (PLAY)

tessellations

in

3-space

MATERIALS NEEDED

Construction paper or cardboard
If Available:
Models of 6 congruent square pyramids fitting
together to make a cube
Models of any pyramids and prisms

There are some objects we could pile together to
fill up a room, and there are some utterly useless
for this purpose. Some of these useless objects are
rocks, footballs, shoes and lamps. They are useless
because they would not fit together perfectly so
that they would fill every point in space. The object
of this chapter is to find those objects we could
use to fill up space—and thus use as a basis for
measuring volume.

Visualize a football and you will see that the shape
is wrong for a third football to fit in any open space
remaining when two footballs are held close together. (Figure 1)

Let us turn our attention to things that take up
space and do fit together so that all of 3-dimen-
sional space can be filled. In geometrical terms, we
want to find some solids that form tessellations in 3-
space. A tessellation in 3-space must, like that in
2-space, fill all the space.

Numerous solids can be used as the sole element in
a tessellation pattern. Cardboard boxes, pads of

FIGURE 1

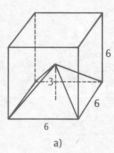

a)

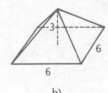

b)

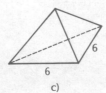

c)

FIGURE 2

rectangular paper, and children's cubical blocks
are possible models. Using your knowledge of tessellations in the plane, you should be able to suggest four or five more solids.

EXERCISE 1:

Start a list of solids, each of which can be used as
the sole element in a 3-space tessellation. If a solid
you think of has as a model a physical thing whose
name is well-known, feel free to use that name in
your list.

On your list no doubt, is the solid cube or building
block. No one can miss at fitting them together
because all six faces are alike. Thinking about a
cube may lead to other solids that will fit together.
Figure 2 attempts to show you one way you can
perceive of the solid cube as made up of other
solids, each of which is thus a basic element in a
tessellation pattern. Consider the triplet of solids in
Figure 2. At the left is a cube with a square pyramid
built inside it, the top of the pyramid being the
center of the cube. By examining this picture you
can see that you could make such a pyramid reaching out from the center of the cube to each of the
six faces. Check it and see if you can visualize six
solid pyramids inside that cube—completely filling
the cube. If so, you rightly surmise that solid square
pyramids of certain proportions will also fill up
space.

EXERCISE 2:

Will any solid square pyramid serve as a basic

element in a tessellation? (A square pyramid is a pyramid built on a square base. A *pyramid* is closed surface, the union of plane surfaces, specifically a base and a set of triangular regions meeting at a point above the base.)

Examination of the square pyramid in Figure 2*b* will help you see another form used in tessellations. You can slice through the peak and down through the diagonal of the square base to make two solid triangular pyramids (two solid tetrahedrons). (The right front one is displayed in Figure 2*c*). Since this is so, the cube can be filled with 12 solid tetrahedrons and that shape of tetrahedron can be used for tessellations of space.

Tetrahedrons differ considerably. The tetrahedron we just obtained by halving a square pyramid had an isosceles triangle for a base and had one face perpendicular to the base.

EXERCISE 3:

a)   Consider a tetrahedron having two faces perpendicular to its base. Will the solid triangular pyramid in Figure 3 serve as the sole element in a tessellation pattern? Why?

b)   Do you think that any solid triangular pyramid will serve as the basis for a tessellation?

EXERCISE 4:

Can a tessellation of 3-space be made using just the solid triangular prism pictured in Figure 4? Ex-

plain. Is your answer regarding Figure 4 dependent on the shape of that prism's triangular base? (A *triangular prism* is a prism built on a triangular region for a base. A *prism* is a closed surface composed of plane surfaces, two bases that are parallel to each other and a set of parallelogram faces connecting these bases. Before 1965, buildings and cabinets were rough models of prisms.)

Will pentagonal prisms fill 3-space? Will hexagonal prisms? A bit of honeycomb is sufficient to convince anyone that hexagonal prisms form a tessellation of 3-space. Look with the artist's x-ray vision at the honeycomb structure in Figure 5.

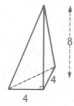

8

4

4

Triangular Pyramid

FIGURE 3

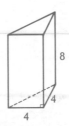

8

4

4

Triangular Prism

FIGURE 4

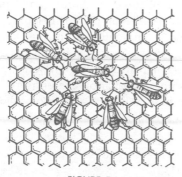

FIGURE 5

Figures 6–8 depict some interesting shapes to consider as sole elements in tessellations patterns. These demand heightened visualization and skill in drawing or making scale models. The class members can apportion these solids amongst themselves as home projects. Each individual should strive to affirm or deny that one of these can be used to fill space in a tessellation pattern. (See Exercise 5.)

The barn in Figure 6 is the union of three types of prisms. (We shall ignore the overhanging piece of roof.) The top is a triangular prism (lying on its side). The section below this in the barn is a trapezoidal prism, and the bottom section is a rectangular prism. We know that the rectangular prism will tessellate space, as might each of the other sections when thought of individually. But what of the barn as a whole? Can it be used as the sole element in a tessellation pattern?

Are the solids pictured in Figures 7 and 8 possible elements for tessellation patterns? Figure 7 shows a solid formed of a triangular prism, a rectangular prism, and an isosceles triangular prism. Figure 8 is a solid rectangular pyramid, different and tricky because its point is so far off center from the rectangular base. Will such a lopsided solid fit with its duplicates to fill all space in a tessellation pattern?

EXERCISE 5:

As mentioned before, choose a solid from the set, Figures 6 through 8. Convince a group whether or not the solid you chose forms a tessellation pattern.

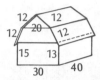

FIGURE 6

FIGURE 7

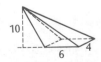

FIGURE 8

You can easily convince others by making models out of cardboard or clay. Use your ingenuity to save yourself some effort and still make an airtight argument. You need not make many duplicates of models if you judiciously use traces on the floor and wall. Use of any results from this chapter and previous ones is commendable, for that is what mathematics is all about.

Having found several solids that can successfully be used in 3-dimensional tessellations, we have many choices of a basic measuring unit in calculating volume. (Your list started in Exercise 1 should have grown.) A volume is measured as a number of solid things that form a tessellation of space; we can report the volume of a room, for instance, as six identical triangular pyramids. From our list, we can see that the solids counted when measuring volume need not be solid cubes unless we feel terribly conventional. The cube is man's traditional choice of a measuring unit. Even using a cube, the calculating of volume becomes a detective game in its own right because a cube seldom sits whole in the container whose volume is being measured. We will wait until Chapter 14 to consider strategies in playing this detective game of finding volume.

1. Cardboard box.

   Artgum eraser.

   Child's building block.

2. A solid square pyramid having its tip above the center of
    the square certainly will work. A
   pyramid such as the one pictured
   here generally will not serve as
   basis of a tessellation pattern.

3. a)   Yes, Figure 3 will serve because of its perpendicular
        faces.

   b)   No, not any triangular pyramid will serve.

4. Yes, because they will pile up from the floor endlessly and
   their bottoms fit together to fill up the plane.

   No, because any triangle will tessellate to fill the plane.

BIBLIOGRAPHY

Ballard, William R. *Geometry*. Philadelphia: W. B. Saunders Co.,
1970.

Johnson, Donovan A. "Paper Folding for the Mathematics
Class." *National Council of Teachers of Mathematics*.
Washington, D.C., 1957.

Schaaf, William L. *Basic Concepts of Elementary Mathematics*.
New York: John Wiley & Sons, Inc., 1966.

Sharp, Evelyn. *A Parent's Guide to the New Mathematics*. New
York: Simon & Schuster, Inc., Pocket Books, Inc., 1964.

School Mathematics Study Group. "Intuitive Geometry."
Studies in Mathematics. Vol. 7. Palo Alto, Calif.: Leland
Stanford Jr. University, 1961, pp. 135–45.

Wenninger, Magnus J. "Polyhedron Models for the Classroom."
National Council of Teachers of Mathematics. Washington,
D.C., 1966.

chapter 8

measure
&
related stuff

MATERIALS NEEDED:

Compass and straightedge and at least one protractor

This chapter should hold something new for each person. Because of the thoroughness of education on the topic of measure, you may find only a few exercises that demand your thought. The chapter is arranged with most exercises at the end so that you can breeze right through all those sections that are merely review. We will look into measure of length, measure of angle, the circle as a set of points, and the concept of tangency to a curve.

8.1 MEASURING LENGTH OF A CURVE

Measure is the assignment of a number to a set of points to indicate some specific aspect of "size." The *measure* of a piece of a line, for instance, involves assigning a number to that piece of line to indicate how long it is. (Score-keeping in tessellations on a line, really!) The "size" of any curve, its measure, is exclusively measure of length. The selection of a unit of measure is influenced by tradition, convenience, and communicability. However, this selection is really arbitrary. One pair of authors indicates this arbitrariness by introducing a unit, called one "handy," and asking children to affirm that the distance from house to factory in an illustration is indeed a little over five "handys." The handy unit for neighborhood shopping trips in the

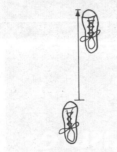

a) The Stride

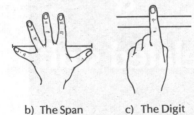

b) The Span   c) The Digit

FIGURE 1

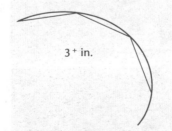

3⁺ in.

a) Use of the Inch to Approximate Length

FIGURE 2

city is usually the length of a city block. For a trip from Kansas City to Phoenix, the mile or kilometer is a more appropriate unit.[1]

The measure of a line segment, $\overline{AB}$, is synonymously referred to as the *length of $\overline{AB}$*, and as the *distance between A and B*. The measure of $\overline{AB}$ is written $m(\overline{AB})$. Use of the symbol $m(\ )$ is reserved for measure. The line segment, $\overline{AB}$, is said to be *bisected* at point P if $m(\overline{AP}) = m(\overline{PB})$. Lengths in the classroom can be measured conveniently by strides, spans, or digits as in Figure 1. You will notice the lack of uniformity in the measure if a group of you compare such measurements, but the comparison will motivate you to establish a standard unit that does not vary with an individual's build.

Happily, not all curves that have to be measured are straight or in straight segments. That would be a bore. Ingenuity in experimentation leads people to clever estimates of such lengths. The tape measure used in sewing follows a physical model of a curve, yet close approximations to that curve's length can be found by using small line segments of one's choice, as is done in Figure 2. The smaller the segments used, the better the approximation.

Basic to the concept of measure of two objects is the idea that the length indicated by a measuring device is not changed when it is moved from one

---

[1] The author hopes the day soon comes when the United States and Canada join other nations in using the eminently convenient metric system in which the kilometer measures cross-country trips.)

object to the other. Man assumes that the compass does not indicate a different length between its points just because we pick it up and move it to a different place. We use the word "congruent" to indicate that length is preserved. We say that a line segment marked off by the compass one place is congruent to a line segment marked off after the compass is moved to another place without changing its opening. Without the assumption of congruence, that movement preserves length, or some other related mathematical concept, all discussions based on length must necessarily be intuitive. We will take the idea of congruence of objects under movement as an assumption in this text; congruence will be used as a basic assumption from which we define other concepts. (See Chapter 11 for more on the idea of congruence and its properties.)

8.2 PARALLEL AND PERPENDICULAR LINES AND SYMMETRY DEFINED

Intuitively, you know something about parallel and perpendicular. You know from having played with boxes and cubes that two opposite plane pieces are parallel and so are four opposite edges. You also know from experience that adjoining sides of a shoe box illustrate perpendicular planes, and that the three edges that meet in a shoe box are mutually perpendicular. We can better define these words, *parallel* and *perpendicular*, now that we have a common vocabulary about length.

Two lines drawn in the plane either intersect or

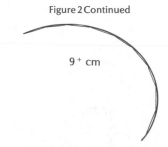

Figure 2 Continued

9⁺ cm

b) Use of the Centimeter to Approximate Length (2.54 cm = 1 in.)

FIGURE 3

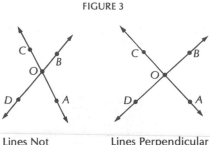

Lines Not Perpendicular Lines Perpendicular

FIGURE 4

they do not. In the geometry Euclid had in mind— the only geometry discussed in this text until the last chapter—two lines lying in a plane are called *parallel* if they do not intersect. Much of modern geometrical research was stimulated by Euclid's fifth postulate about parallel lines: "Through a point not on a line, there is exactly one line parallel to the given line." (Ironically this stimulation came from the efforts of some to show, mistakenly, that the fifth postulate was redundant and could be proved from the other axioms; This effort initiated the major mathematical discovery of the 19th century—that other geometries are formed by substituting other possibilities for the fifth postulate.)

Two intersecting lines are defined as perpendicular if the four rays centered out from the point of intersection form congruent adjacent angles. (After we have discussed properties of congruency in Chapter 11, you will see that, to show two lines perpendicular, it is sufficient to show that any pair of adjacent angles are congruent.) Notice the adjacent angles, $\angle AOB$, $\angle BOC$, $\angle COD$, and $\angle DOA$, in both parts of Figure 3. Since two things are congruent if the points match after one is picked up and moved on top of the other, $\angle AOB$ in the right-hand picture is congruent to both its adjacent angles, $\angle BOC$ and $\angle DOA$.

The concept of perpendicularity was the basis of an ancient version of the level. This level was in the shape of an isosceles triangle with a plumb line hanging down from the vertex. When the line hung in front of its exact center, the base was horizontal. (See Figure 4.)

Another application of the concept of perpendicularity is seen in symmetry about a line and symmetry about a plane. Now that "perpendicular lines" and "measure of length" are usable phrases, it is possible to define symmetry. Remember how in Chapter 2 we used mirrors and paper folding to determine whether or not sets of points were symmetrical? Folding to test for symmetry is a physical test of the possibility that somewhere there is a line across which lie paired points that are set in this manner: (1) at equal distances from the line (fold) and, (2) on a line perpendicular to the fold.

There are two kinds of symmetry possible for points in a plane—symmetry with respect to a point and symmetry with respect to a line. Two points are said to be *symmetrical with respect to a point*, P, if P bisects the line segment joining the two points. Two points are said to be *symmetrical with respect to a line* if that line is the perpendicular bisector of the line segment joining the two points. (A line is the *perpendicular bisector* of a line segment if it bisects the line segment and is perpendicular to it.) Whole sets of points can be considered symmetric with respect to a point (line) when each pair of points is symmetrical. Two figures can be considered symmetrical with respect to a point (line) if each point in one figure has a symmetrical point in the other drawing. Now that you have formal definitions of the two symmetries in each plane, it would be wise to turn back to Figure 13 in Chapter 2 to check these definitions of the point and line symmetries pictured there.

For points in space, there are three kinds of sym-

FIGURE 5

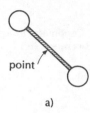

point

a)

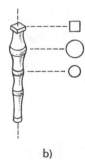

b)

metries possible. The definitions of point and line symmetry are worded exactly the same as the two just discussed, but you may find them an effort to conceptualize without some additional discussion. Both point and line symmetry in space entail an extension in all directions of the comparable symmetry seen in two dimensions. For example, a dumbbell has a point of symmetry. To affirm this fact with the help of Figure 5a, we mentally check for a property of point symmetry in space, a point symmetry in each plane passing through this proposed point of symmetry. Sets of points symmetrical about a point P will have symmetry about P in any place containing P.

A table leg cut on a lathe has a line of symmetry. In examining this table leg, (pictured in Figure 5b) we see that line symmetry in 3-dimensions implies the existence of line symmetry in each plane containing the axis of symmetry, and thus implies point symmetry in each plane perpendicular with respect to the axis of symmetry. Affirm this as you examine the picture of the table leg and several of its plane sections as shown in Figure 5b. Line symmetry is the ideal in many artifacts made to balance and in those manufactured by revolution as on a potter's wheel.

EXERCISE 1:

a)   Name or sketch four items that should have line symmetry.

b)   Name or sketch two objects that should have point symmetry.

c) What is the simplest way to sketch a table leg?

In space, we have a third kind of symmetry, discussed from an intuitive approach in Chapter 2. It is called symmetry about a plane, the reflection effect noted between a right-hand glove and its reflection—and thus between right-hand and left-hand gloves placed thumbs parallel. We can define two sets of points as symmetrical about a plane if pairs of points are symmetrical about a plane. Two points are *symmetrical about a plane* if the plane passes through the midpoint of the line segment connecting the two points and is perpendicular to that line segment.

8.3 CIRCLE AND ANGLE MEASURE

Most of you know what a circle looks like and can suggest one by spinning a weight through the air on the end of a string or by tracing around a tin can. While spinning, the metal weight passes through a set of points ideally all the same distance from the junction of the hand and the string. This suggests the formal definition of a circle, perhaps the easiest definition in all of geometry. The *circle* is a special simple closed curve, defined as the set of all points in a plane at a given distance, *r*, from some point, *C*, in that plane. The point C is called the *center*, the distance *r*, the *radius*, and 2*r*, the *diameter*. You use this definition of a circle when you describe the orbit of a weight spinning, and you used it when you followed a workbook to trace out

a)

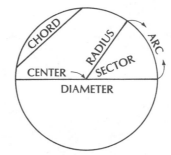

b) Vocabulary Associated with a Circle

FIGURE 6

a circular cylinder by moving a pencil tied on a string. (See Chapter 2.)

It is somewhat easier and more accurate to use, instead of a string held inaccurately by a finger, a metal gadget with a point at one end and a pencil clamped onto the other end. This easily adjustable dime store item is called a compass. (See Figure 6a.) Play with one if you haven't before; you can make many beautiful designs of interlocking pieces of circles. Pieces of circles, incidentally, are called *arcs* and the region bounded by an arc and two radii of a circle is called a *sector of the circle*. (See Figure 6b.)

In examining Figure 6, we see correspondence between the length of the circular arc, the amount of angle opening (perceived) between the two radii cutting off that arc, and the length of the chord connecting the two points where the radii intersect the circle. You can see that if we were to lengthen the arc, it would increase both the opening of the angle and the length of the corresponding chord. We will use the length of the arc as the basic measure of an angle.

Let's drop back and gather our data so that you can see why the author chooses the length of a circular arc as basic to defining angle measure. First of all, an angle is a union of two rays having a common starting point. It is a special curve whose length is of no interest to us for it is of infinite length. (We can, however, get an interesting measure on the amount of opening between the rays.) Secondly, there is one use of the line segment that might make sense in measuring an angle. The more opening an angle has, the longer the toothpick (model for a line segment) needed to prop it open if the ends of a toothpick were placed 1 inch out on both rays. See Figure 7, where the toothpicks are of the approximate length $\frac{1}{2}$ inch, 1 inch, and 2 inches.

But this measure of an angle is meaningless for the straight angle (there is one pictured in Figure 8a) and for angles even bigger than the straight angle. (Figure 8b) (Notice that angles are traditionally measured counterclockwise.)

Here is another defect in the measuring of angles by their chords. We want the measure of $\angle AOC$ to be expressible as the number of smaller angles that will fit inside. Yet Figure 9 shows that the chord measure of the union of two adjacent angles is less than the sum of the measures, another definitely undesirable characteristic. We must look for another unit with which to measure an angle.

A more useful way to perceive of an angle is to see it as centered in a circle with both its rays starting at center and intersecting the circle at two points, cutting off an arc. This allows us to associate an arc with each angle, even with the straight angle. We can associate the special arc called the *semicircle* with a straight angle.

EXERCISE 2:

(Exploring distance around a circle). Cut out a cir-

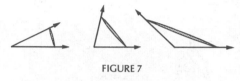

FIGURE 7

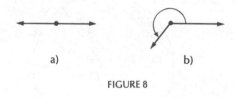

a)   b)

FIGURE 8

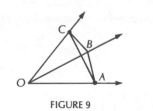

FIGURE 9

cular patch of thick cardboard. Knot a piece of string so that the distance between the knots is that of the radius of the cutout circle. Use the string to mark arcs of the same length around the circle. Approximately how many radii lengths is the distance around your circle?

EXERCISE 3:

Bring in a bicycle or a wheel from a bicycle or wagon. Cut a piece of string the length of the radius of your wheel (radius is half the diameter). Use your piece of string to mark off consecutive radii along a straight line on the floor or sidewalk. Now put your wheel on the line at the first mark and chalk the tire where it touches that mark, then slowly roll the wheel along the line and mark the spot on the floor where the chalk mark returns to the floor.

a)   Will the distance between the touchings of the tire mark be the same for another roll of the wheel?

b)   Is the distance measured along the floor connected with any particular distance on the wheel?

c)   What is your estimate of the number of radii measured on the floor between where you started rolling and where the chalk mark again hit the floor?

d)   Is there any connection between this number and the number of radii estimated in Exercise 2?

Noting the results of this play with circular objects, we see that the distance around any circle is the same if we measure in number of radii. So $\frac{1}{8}$ turn about a circle can be given in the same units, regardless of the size of the circle. See Figure 10 for the obvious advantage of measuring arcs as parts of the circle circumference, and for the obvious advantage of arc measurement over chord measurement.

The distance around any circle appears to be about 6.28 times the length of the radius of the circle. It is exactly $2\pi r$ units around a circle of radius r units.

Given a circle of radius r, we can define angle measure so that there is just one length of arc for each angle. If there is a straight angle, then we could report its measure as πr, the length of a semicircle. If an angle is of measure greater than one revolution, then its corresponding arc is of length greater than $2\pi r$, the distance around the circle. For our purposes negative angles are ignored.[2]

We name a unit of angle measure called the _radian_; we say that an angle measures _K radians_ when the portion of the circle cut off by the angle measures K radii in length. We shall define the measure of an angle to be the length of the circular arc associated with that angle. There are π (approximately 3.1416) radii needed to reach around a semicircle, so we say that the straight angle has a measure of π radians and that half of a straight angle (a right angle) has a measure of $\pi/2$ radians and so forth.

Another popular unit used in communicating angle

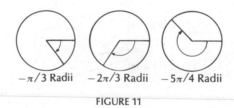

Each Arc is $\frac{1}{8}$ of its Circle's Circumference, $\pi/4$ Radii.

FIGURE 10

$-\pi/3$ Radii $-2\pi/3$ Radii $-5\pi/4$ Radii

FIGURE 11

measurement is the degree. This is a narrow angle chosen by man so that 180 of them is the measure of the straight angle; from this it follows that the right angle has a measure of 90 degrees (90°), and the angle that is $\frac{1}{3}$ of a straight angle has a measure of 60 degrees. Although the degree type of measure is certainly not the only type of measure, it is desirable to be ready to communicate in this very popular language. You can think of degree measure as a dividing up of the circle into 360 angles or arcs, each measuring 1 degree. The measure of any size angle (or arc) can be stated in degrees by a quick conversion:

The measure of $\frac{1}{4}$ of a straight angle is $\frac{1}{4}$ times $180° = 45°$

Written in mathematical shorthand, $m\left(\dfrac{\pi r}{4}\right) =$

$\frac{1}{4} \cdot 180° = 45°$

Comparing two angles to find out which is the larger can be facilitated by using a transparency, a compass, or a piece of folded paper. (The compass retains chord length. The paper folded into a wedge to fit one angle retains that angle's measure and fits entirely within a second angle that is bigger. See Figure 12.)

A more refined process of comparing the sizes of angles involves assigning numbers indicative of size to both angles. Obviously, definite measures written down are an advantage. You can then compare numbers and conclude that an angle measuring $5\pi/4$ radians, for instance, is greater than one measuring $5\pi/6$. And an angle measuring 31° is just a

FIGURE 12

little greater than one measuring 29°. A tool to measure angles can easily be made or bought. It is called a protractor. It looks like a ruler with a semicircular band glued to it. All that really matters is that the center be marked along the protractor's straight edge and that the semicircular band be marked in some units of arc length. (Figure 13)

All these methods and many more help you compare sizes of angles in the plane. What we have discussed will also apply to measuring an angle between two intersecting planes.

8.4 MEASURE OF AN ANGLE BETWEEN TWO PLANES

Let us take a look at the possibilities for relations between lines and planes in space before we get involved in the concepts and the vocabulary of planar angles. A set of points is said to be parallel to a plane if their intersection is empty. Thus a line can be parallel to a plane and two planes can be parallel.

EXERCISE 4:

(An exercise on exploring the intersection of various sets of points.)

a)   Consider all the possible intersection sets for a line and a plane. Describe each situation verbally and make a sketch demonstrating each comment.

b)   How many planes can contain a given line?

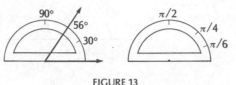

FIGURE 13

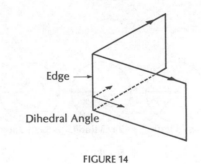

Edge ⟶

Dihedral Angle

FIGURE 14

c)   When two distinct planes have a nonempty intersection, this intersection set of points can be described verbally; do this and make a supporting sketch of two planes intersecting.

Let the two flaps of a manilla folder represent planes. We can see that they intersect in a line (as do any two distinct planes). We can measure the angle between the two planes, called a _dihedral angle_, in much the same way as we measured angles in the plane. Opening the top of the folder counterclockwise, we see a semicircle traced out by any point on the moving flap. To indicate a particular position of the top flap in relation to the one on the desk, we can indicate a particular position on the semicircle—and thus a particular dihedral angle between the half planes. The dihedral angle measures a few degrees when we first open the manilla folder counterclockwise, and measures 180° when it lies flat on the desk.

We follow now with the formal definitions of dihedral angle and its measure. The union of two intersecting half planes forms a _dihedral angle_ (see Figure 14). The line of intersection of the half planes is called the edge of the angle. The measure of such an angle builds on what we already know. Draw a ray in each half plane, with each ray perpendicular to the edge at a point on the edge. The angle between these rays is a _plane angle_ associated with the dihedral angle. All plane angles of a given dihedral angle are congruent.

Figure 14 shows several plane angles of a dihedral

angle. The measure of the dihedral angle is defined as the measure of any of its plane angles. It follows that two planes are perpendicular if any dihedral angle there has a right angle for its plane angle.

EXERCISE 5:

a)   Open a book to form a dihedral angle of 30°. Can you check your guess with a protractor?

b)   If the ends could not be seen, as in a cabinet, how might you check a dihedral angle for accuracy? For instance, pretend that you asked a cabinet-maker to build a window seat for an alcove in your condominium. The cross-section is pictured at right. What could you do in the cabinetmaker's shop to be sure the 60° angle extended all the way to the floor, assuming you can get your arm in?

8.5   TANGENT TO A CURVE

At the moment of skid, a car losing its grip on the road goes out of control—in a path tangent to the road at that point. If you haven't had the dubious pleasure of seeing a car slide off a road, don't plan it as an experiment. Instead, perform a physical experiment or a thought experiment with a bean bag on the end of a cord: You are spinning wildly in the center of a gymnasium, whirling the bean bag around you in a great circle on the floor. Suddenly the cord breaks. Watch the straight line path of the bean bag as it speeds away from you across the floor. That path is the tangent line to the circle

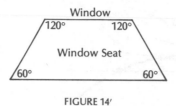

Window

120°   120°

Window Seat

60°   60°

FIGURE 14'

FIGURE 15

Trajectories

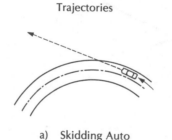

a)   Skidding Auto

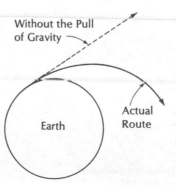

Without the Pull
of Gravity

Earth

Actual
Route

b)   Launched Space Ship

at the point where the bean bag was last on the circle. (Same for the auto in Figure 15a.)

You are fortunate if you have seen this straight path of an object suddenly unrestrained by centripetal force, for many such occurrences in nature show curved paths modified by gravitational pull. (See Figure 15b for the usual route of a space ship that rockets from earth's soil.)

The continuous part of any curve has tangents at each point along the curve. The definition of tangent drawn to a point involves, amongst other things, agreement between left and right as to the prevailing direction of curvature; hence, there is no tangent line at isolated points, end points, or points where the curve changes abruptly. If you wish, glance ahead to Figure 17 to check your intuition as to where tangent lines are undefined and how they look when they are defined.

The definition of a tangent line drawn to a point P on a curve can be made quite concise. Consider the meaning of the tangent line at P by looking at Figure 16a. Look at the line through the points P and Q, where Q is a point to the right of P along the curve. $\vec{PQ}$ is called a secant line because it connects two points on a curve. Consider the succession of secant lines through P and points on the curve ever closer to P (such as R and S in Figure 16a). The _tangent line_ at P is defined as the limit of secant lines through P as the other point used approaches P—providing that the limit from the left is the same. Can you see the tangent line at P as a

FIGURE 16

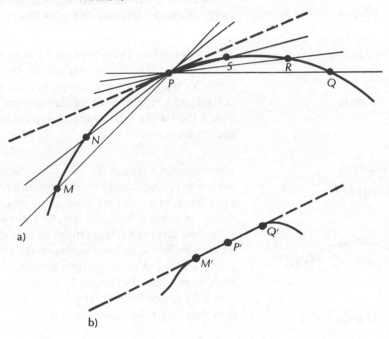

a)

b)

FIGURE 17

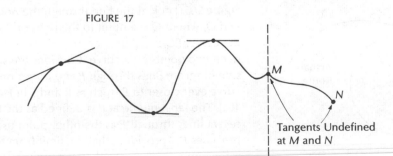

Tangents Undefined
at M and N

limit line in Figure 16a? (It is the dotted line.)

Can you likewise see that there is a tangent at P' (dotted line in Figure 16b), though there is no curvature around P'?

You can now see why the tangent lines slant as they do in Figure 17, and you can understand why the tangent line is undefined at M and N.

Many of us, in our youth, heard someone tell about the tangent to a circle. The circle is a special curve having the same amount of curvature at each point. For this reason, the tangent drawn to a point on a circle has some special properties.

If you were to draw a line through a point P on a circle so that the line passed through no other point of the circle, that line would be unique and it would be the tangent line. You should check this uniqueness at point P on the circle in Figure 18 by drawing a line that touches the circle only at P and then rotating your ruler around P, checking for any other lines that intersect just at P.

For a circle, it is sufficient to define the tangent at P as the unique line intersecting the circle in only point P. This one line through P also has a special relationship to the radius drawn to point P. The unrestrained bean bag and the spaceship unaffected by gravity trace paths tangent to the circular path they previously followed. These physical examples help us to see that the tangent to a circle is perpendicular to the radius drawn to point P. Sketch the radius to P in Figure 18 and measure the angle between the two lines with a protractor or a piece of folded paper, affirming that the angles are right angles.

Circles tangent to lines have interesting properties—you can enjoy a few of them in the exercises. Many other curves have especially well-behaved tangent lines, yet the greatest benefit is reaped by those who understand the concept of tangent well enough to extend the concept to points on any curve. This extension furnishes a useful intuitive interpretation of the derivative in calculus[2] and has a practical application in the calculation of gradient for roads.

You are not only making a worthy contribution to your own education when you learn what is meant by the term "gradient"; you are also seeing a practical application for finding a tangent to a point on a curve. The engineer needs to know the measure of the angle between the horizontal and a curved outline of a hill. No outline of a road going up a hill is a sudden upward slanting line segment, for that jerk would be too hard on the vehicle. Instead each hill is a combination of concave and convex curves, as exaggerated in Figure 19a. If we were to sketch tangents at each point on this outline of a hill, there would be one that was steepest. The ray along this steepest tangent line and a horizontal ray determine an angle. Sixty percent gradient, as pictured in Figure 19a, means that this angle opens 60

---

[2]The derivative at a point P is generally based on the inclination of the line tangent to the curve at P.

units vertically for every 100 units horizontally. Six percent gradient, a feasible highway gradient, is isolated and pictured in Figure 19b.

FIGURE 18

FIGURE 19

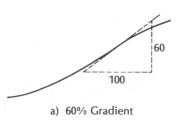

a) 60% Gradient

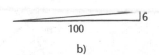

b)

8.6 EXERCISES FOR CHAPTER 8

6. Draw one straight line on a whole sheet of paper and then draw some circles that have this line for a tangent. Considering the subset of all the circles tangent at one point on this line, what can you say about their centers?

7. Draw two parallel lines on a whole sheet of paper and then draw circles tangent to both lines. What can you say about their centers?

8. a) Draw two intersecting lines, r and s, intersecting near the middle of a whole sheet of paper. Sketch some circles, if any, that are tangent to both lines. Can you draw at least one in each of the four sections into which the paper is separated?

b) What can you say about the centers of all circles in one section of the paper?

c) Let O be the point of intersection of r and s. Let R be the point of intersection of one circle and line r, while S marks the point of intersection of that same circle with line s. Compare the lengths of $\overline{OS}$ and $\overline{OR}$. What do you notice about the lengths of the line segments drawn from O to the two points of tangency of a circle?

Figure 20a shows two parallel lines and a line intersecting both of them. Use a protractor to measure one of the two angles above the top line and compare that to the measurement of a similarly placed angle above the bottom line. If the lines are truly parallel, the angle measures should be equal for these corresponding angles. This is true no matter what the position of the line crossing the parallel lines. (Reexamine the gradient drawing as it appears in Figure 20b, and notice that the tangent to the slope intersects a set of horizontal parallel lines. It is because the angle measures are equal that it does not matter where the horizontal ray is drawn to calculate gradient.) Henceforth, feel free to build on the following statement as if it were a fact. Here is the fact of this geometry: Two lines are parallel if, and only if, they form equal angles when crossed by a _transversal._

9. Use this idea to finish Figure 20c by constructing a line parallel to line m and passing through point P, if I tell you that line s is perpendicular to line m. Explain what you did to construct a line parallel to m.

10. There is a theorem that a pair of lines in the plane are parallel if, and only if, the alternate interior angles are equal. We can prove it as an exercise taken in two parts. Use the fact that the measure of a straight angle is π radians.

    a) Make an argument for equality of the measures of ∠1 and ∠1′ in Figure 21, assuming that the two lines are parallel.

    b) Can you argue the converse, that the lines

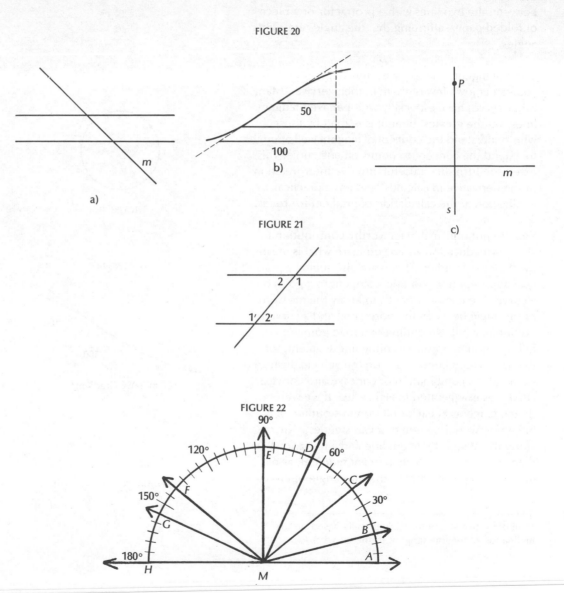

FIGURE 20

a)

b)

c)

FIGURE 21

FIGURE 22

must be parallel if $m(\angle 1) = m(\angle 1')$ and $m(\angle 2)$ $= m(\angle 2')$?

11. Using what you discovered in Chapter 4 about the total number of angle degrees around any point and the total angle measure of a quadri-lateral, and using anything in the text so far, can you make an argument proving that, in Figure 21, $m(\angle 1) + m(\angle 2') = 180°$ when the two lines are parallel. (Remember, we have already discussed how a line perpendicular to both parallel lines forms angles whose mea-sures add to 180°, and this is helpful if you choose to construct an additional line segment to make a quadrilateral.)

12. The word "radian" as a unit of measure sug-gests relative distance around the circle. A piece of string as long as the radius will cover one radian of arc when draped around the circle.

    a) If the circle has diameter 4 inches, how long is a semicircle? How long is 1 radian of arc?

    b) If a circle has a radius of 1 centimeter, how long is 1 radian of arc?

    c) As the circle gets smaller, does 1 radian of arc cut off smaller, larger, or equal angles?

13. Figure 22 is schema of a protractor marked at intervals of 5°. The center is at M. Rays with endpoints at M are lettered. Complete the following sentences, estimating to the nearest degree.

    a) the measure of $\angle FMG$ = _____

    b) $m(\angle AMB)$ = _____
       $m(\angle AMG)$ = _____
       $m(\angle FMH)$ = _____

    c) Name an angle whose measure is approximately 42° _____
       23° _____
       90° _____

    d) $m(\angle DMF) + m(\angle FMG)$ = _____

14. *Skew lines* are lines that neither intersect nor are parallel, like the lower edge of a wall and the upper edge of an adjacent wall in a room. Thus it follows that skew lines do not lie in the same plane, for if they did, they would be parallel or intersect.

    a) Define such a thing as *skew planes*, or declare them definitely nonexistent.

    b) Sketch or mention by name two familiar examples of pairs of skew lines. (Look to the shoe box for ideas if you feel stymied).

    c) Mention by name three examples of sets of at least four parallel lines.

    d) In the following situation, is line $\overleftrightarrow{AB}$ neces-sarily perpendicular to plane n? Planes m and n are parallel; point A lies in plane m and point B lies in plane n; line $\overleftrightarrow{AB}$ is perpendi-cular to plane m. (Line $\overleftrightarrow{AB}$ is said to be per-pendicular to a plane if it is perpendicular to any pair of lines intersecting $\overleftrightarrow{AB}$ and lying in the same plane). Experiment with models, or try reasoning indirectly in a manner begin-ning like this:

Suppose $\overleftrightarrow{AB}$ is not perpendicular to n.
Then $\overleftrightarrow{AB}$ is not perpendicular to some $\overleftrightarrow{BC}$
lying in n.

.
.
.
.

15. Fold a sheet of paper twice and cut off a corner of your folded sheet. Unfold and examine
your region of the plane for lines of symmetry.
Mark each line of symmetry with a pencil, and
bring it along to show others.

16. Examine each of the sets of points pictured at
right and decide whether there is symmetry
about a line, a point, or none at all. If you
have found any symmetry, place your line
and/or point of symmetry in the picture. Remember, several lines of symmetry intersecting
in one point may mean that there is also a
point of symmetry.

17. A man builds forms for a concrete floor. He
makes one side of the form sloping to allow
for water drainage while his friend makes the
opposite side of the rectangular form level.

a) Explain how he could use some form of
sighting to detect that the top edges of the
opposite forms are not parallel.

b) If they pour the concrete and drag a
board across the frame so that the concrete is
even with the top of the forms, will the surface
be a plane surface and will it drain?

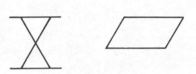

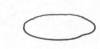

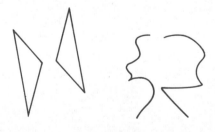

FIGURE 23

18. a) Through a given point 3 feet above a floor,
how many planes can pass that are parallel to
the floor?

b) Through a given point 3 feet above a floor,
how many lines can pass that are parallel to the
floor?

c) Through a given point 3 feet above a floor,
how many lines can pass that are parallel to a
yardstick lying on the floor?

19. If a line intersects one of three parallel planes,
must it intersect the others?

20. Consider planes p, q, and r. When, if ever, can
$p \cap q \cap r$ (the intersection of the three planes)
be

a) one line?

b) two lines?

c) three lines?

d) four lines?

e) empty (no point of intersection)?

f) one point?

21. The perpendicular bisector of a line segment
might turn out to be different if we don't confine the perpendicular bisector to a plane in
which the segment lies. Visualize the set of all
points equidistant from both endpoints of a
pencil, for that is the perpendicular bisector
of the pencil. Describe this set of points.

1. a)   a yardstick, a softball, a book, a wastebasket.

    b)   a bicycle wheel, a yardstick, and a softball.

    c)   The simplest way is to draw half of a table leg and then to fold the paper on the leg's line of symmetry and trace the outline so that the table leg looks like a real one cut on a lathe.

3. a)   Yes.

    b)   The circumference of the bike wheel is the same as the distance marked along the floor.

    d)   The number of the radii around the circle is the same for both the bicycle wheel and the cardboard circle of Exercise 2.

4. a)   You should have three pictures with the following verbal descriptions: No point in common, one point in common, an infinite number of points in common.

    b)   An infinite number of planes contain a certain line.

    c)   Two distinct planes intersect in a line.

5. a)   Not unless you squinted and assumed uniformity of the opening, a bad assumption.

    b)   If I were building a cabinet, I'd slide back and forth a piece of wood cut on a 60° angle and watch that it just barely fit everywhere.

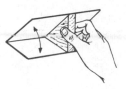

6. All circles tangent at a point have their centers on a line, the line passing through the point and perpendicular to the line of tangency.

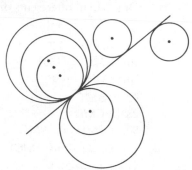

7. The line of centers is parallel to both lines and lies halfway between.

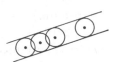

8. Partial answer.

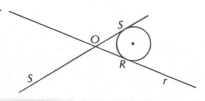

9. To make right angles at P, construct a line through P perpendicular to line s. Now both the new line and m make right angles where crossed by transversals, hence are parallel.

10. a) $m(\angle 1) = m(\angle A)$, where $\angle A$ is opposite $\angle 1'$.

     Because $m(\angle 1') + m(\angle 2') = \pi$ radians

     and $m(\angle A) + m(\angle 2') = \pi$ radians,
     $$m(\angle A) = m(\angle 1')$$

     Hence $m(\angle 1) = m(\angle 1')$.

12. a) 2π inches; 1 radian or arc is 2 inches long.

b)   One radian is 1 centimeter long.

c)   One radian of arc cuts off angles of equal measure.

13.   a) $m(\angle FMG) = 13°$.

b) $m(\angle AMB) = 15°$.
$m(\angle AMG) = 153°$.
$m(\angle FMH) = 40°$.

c) $42°: \angle AMC$.
$23°: \angle CMD$.
$90°: \angle EMH$ or $\angle AME$.

d) $m(\angle DMF) + m(\angle FMG) = m(\angle DMG)$ or $88°$.

14.   a)   There is no such thing as skew planes; two planes are either parallel or they intersect.

b)   A railroad track on a bridge and an edge of the road the bridge spans.

c)   A music staff exemplifies a set of five parallel lines.

d)   I will not tell you now, but a picture like this should be your starting point.

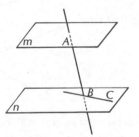

16.   No symmetry for these figures.

Line symmetry for each of the first three figures, the semi-circles, the horseshoe, and the mutated Roman numeral, with several lines of symmetry in the first and third.

Point symmetry for the parallelogram, the double J's, and the double triangles.

Look to see if the above is all that can be said. A figure can have both line and points symmetry.

18.   a)   one plane;   b)   an infinite number of lines;
c)   just one line.

19.   Not if the line lies in the plane it intersects. If the intersection is just one point, the line must intersect the other planes because the line is not parallel to the other two planes and they extend indefinitely.

20.   a)   You can see that occur as you visualize three planes forming different dihedral angles with the same edge.

b)   Cannot be.

e)   If all three planes are parallel.

f)   Yes, if a third plane intersects but does not contain the line of intersection of the other two.

21.   A plane passing through the midpoint of the line segment.

Abbott, Janet S., et al. *Mirror Magic*. Pasadena, Calif.: Franklin Publications, Inc., 1968.

Bentley, Potts. *Geometry, Part I—Discovery by Drawing and Measurement*. Boston: Ginn & Co., pp. 50–51.

Copeland, Richard W. *Mathematics and the Elementary Teacher*. Philadelphia: W. B. Saunders Co., 1966, pp. 264–7, 270–80, 298–304.

Garstens, Helen L., and Jackson, Stanley B. *Mathematics for Elementary School Teachers*. New York: The Macmillan Co., 1967, pp. 285–92, 318.

Horn, Sylvia. *Learning About Measurement*. Pasadena, Calif.: Franklin Publications, Inc., 1968, pp. 5, 15–19.

Johnson, Donovan A. "Paper Folding for the Mathematics Class." *National Council of Teachers of Mathematics*. Washington, D.C., 1957.

McFarland, Dora, and Lewis, Eunice M. *Introduction to Modern Mathematics*. Boston: D. C. Heath Co., 1966, pp. 190–6.

Marks, Smart, and Purdy. *Sets in Geometry*. 2d. ed. Boston: Ginn & Co., 1963, Ch. 7.

Mueller, Francis J. *Arithmetic, Its Structure and Concepts*. Englewood Cliffs, N.J.: Prentice-Hall, Inc., 1964, pp. 314–24, 327–9, 332–3, 338–40.

National Council of Teachers of Mathematics. *Measurement, Topics in Elementary Mathematics for Elementary School Teachers*. Booklet 15, 1968.

Newbury, N. F. "Quantitative Aspects of Science at the Primary Stage." *The Arithmetic Teacher* (December 1967), pp. 641–4.

Nuffield Mathematics Project. *Shape and Size 2*. New York: John Wiley & Sons, Inc., 1968, pp. 79–80.

Nuffield Mathematics Project. *Shape and Size 3*. New York: John Wiley & Sons, Inc., 1968, pp. 11, 29.

Halliday, Helen K., et al. *Sadlier Series Contemporary Mathematics*. New York: William H. Sadlier, Inc., 1968, pp. 118–22, 328–32, 374, 382, 387.

Peterson, John A., and Hashsisaki, Joseph. *Theory of Arithmetic*. New York: John Wiley & Sons, Inc., 1963, pp. 227–31, 256–7.

Schaaf, William L. *Basic Concepts of Elementary Mathematics*. New York: John Wiley & Sons, Inc., 1966, pp. 293–4, 321–7.

School Mathematics Study Group. *Studies in Mathematics*. Vol. 9. Revised Edition. Palo Alto, Calif.: Leland Stanford Jr. University, 1963, pp. 371–2.

School Mathematics Study Group. "Intuitive Geometry." *Studies in Mathematics*. Vol. 7. Palo Alto, Calif.: Leland Stanford University, 1961, pp. 134–45.

Walter, Marion. "An Example of Informal Geometry: Mirror Cards." *The Arithmetic Teacher* (October 1966).

chapter 9 (PLAY)

projections

MATERIALS NEEDED:

Household solids
Flashlight
String and beads
Checker or chess board
Equilateral triangular region
Circular region

Have you ever noticed the shadows thrown on a wall when an object is placed in front of a candle or a flashlight beam? Examination of projections— seen as shadows—allows you to learn some of the properties of projective geometry.

Perhaps you have figured out intuitively what similarities can be expected between an object and its shadow. You would probably look up in consternation if you saw a certain shadow on the ground beneath a suspended bird-feeding ledge that had a rectangular shape. For curiosity's sake, which shadow in Figure 1 would you accept without surprise? Mark the letter S beneath those you might see sometimes, and mark N beneath those you would never see.

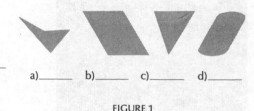

a)_____ b)_____ c)_____ d)_____

FIGURE 1

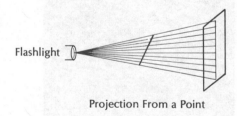

Flashlight

Projection From a Point

FIGURE 2

Only the patch in Figure 1b might you see on the ground near a rectangular ledge.

The sun's rays are almost parallel to each other. Although there are some mathematical similarities between parallel projections (modeled by sunlight) and projections from a point of light, we choose in this chapter, for mathematical reasons, to concentrate on shadows cast by candles, flashlights or slide projectors so that our source of light seems to come from a point. In Figure 2, we see the effect of a flashlight beam on a line segment. If you had trouble visualizing what the shadow would be, put beads on a string at certain intervals and watch the shadow when you hold the string taut between your fingers, modeling a line segment with points on it. Now move the line segment to different positions, noting the shadows of the beads. By so doing you may form some general principles that will help you in the following exercises.

EXERCISE 1:

a)   A chessboard is held between a flashlight and a wall. In Figure 3, write the letter S directly beneath each shadow that could appear on a wall when a flashlight is held behind a chessboard. Write the letter N beneath those shadows you think you would not see.

b)   Do straight edges ever have rounded arcs as shadows?

Are perpendicular edges necessarily perpendicular in the shadow?

Do sides of equal length have shadows of equal length?

EXERCISE 2:

a) Repeat the instructions of Exercise 1a, considering a shadow cast by a patch whose outline is an equilateral triangle instead of a chessboard. Mark the letter S or N on the second line beneath each of the possible shadows in Figure 3.

b) Is the projection of a triangle always a triangle? (i.e., is the property of having three sides preserved by projection?) Explain.

c) Is the length of a side preserved by projection? (i.e., is the property of being x inches long preserved by projection?)

EXERCISE 3:

a) Repeat the instructions of Exercise 1a, considering a circular patch and answering S or N on the bottom row of lines beneath each picture.

b) Does a circular patch always make a circular shadow? Does its shadow always have curved sides?

By looking at the answers and testing your ideas with flashlight shadows you no doubt formed the following concepts:

FIGURE 3

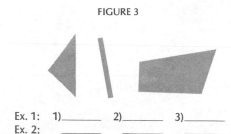

Ex. 1: 1)_____ 2)_____ 3)_____
Ex. 2:   _____   _____   _____
Ex. 3:   _____   _____   _____

Ex. 1: 4)_____ 5)_____ 6)_____
Ex. 2:   _____   _____   _____
Ex. 3:   _____   _____   _____

Ex. 1: 7)_____ 8)_____
Ex. 2:   _____   _____
Ex. 3:   _____   _____

1. A flat patch of any shape can be turned parallel to the light so that only a line segment shows on the wall. This case, called the exceptional case, we shall henceforth omit from general consideration.

2. A curved edge of a model projects a curved image on the wall, and the shadow bulges outward, corresponding to the original. In addition, line segments bounding a model cause line segment boundaries on its shadow, as you no doubt found in experimenting with the strung beads.

3. Distances between points are not maintained; hence, a shadow can have a shorter or longer side than does its model.

4. The number of sides (and arcs) of a patch is maintained (in all but the exceptional case).

We call this last statement a property of projection. We say that projection has the property of preserving the number of sides of a polygonal patch. To see how this happens, let us examine the shadow of a rectangular patch on a wall. In all but the exceptional case, one edge is held higher from the floor than the opposite edge. The higher edge and its two vertices will show in the shadow, and because the rectangle is held in the nontrivial position, the shadow will also show the lower edge and its two vertices. Similarly, for any quadrilateral, all vertices and the connecting edges will show in the projection. This analysis can be applied to patches and solids of any number of curved and straight edges.

By holding a beaded string taut in a flashlight beam, we discovered that a line segment projects onto a line segment. (Notice that projection is more specific than topological equivalence, which equates arcs of any curvature.) We also noted that all points of the line segment appeared in the same order in the shadow; a point between end points of a segment will have a shadow between end points of the segment's shadow. These properties imply that a line segment lying on a patch will be projected wholly within the shadow of that patch. Might this last statement regarding projection of a line segment allow us to infer properties of projection beyond those already discussed in the last paragraph?

We need a new word. A set of points is _convex_ if the set contains the line segment joining any two of its points. All of the regions pictured in Figure 3 are convex except the seventh. Try moving a ruler around and you will see that a line segment connecting two points of a convex region always lies completely within that region.

EXERCISE 4:

a)   Make an "original," a cardboard convex region different from any used so far in this chapter. Sketch this original convex region in the space furthest to the left below, and sketch four shadows of this patch in the four spaces beside it. Label each of the four shadows as convex or not convex.

b)   Cut out a nonconvex region different from any used so far in this chapter. Sketch the original and four shadows, and label them as convex or not.

Convex ⎯⎯⎯   ⎯⎯⎯⎯   ⎯⎯⎯⎯   ⎯⎯⎯⎯   ⎯⎯⎯⎯

(my original)

Non-convex ⎯⎯⎯   ⎯⎯⎯⎯   ⎯⎯⎯⎯   ⎯⎯⎯⎯

(my original)

c)   Can you see any reason why all projections of convex regions would be convex—or not so?

Will all projections of nonconvex regions be nonconvex? State your reasons.

By the way, the term _convex_ refers to solids and curves as well as to regions. Look anew at solids such as hassocks, smooth candles, baseballs and the like, and you will discover that the word convex also applies to solids modeled by such objects. Look back at the definition to affirm this and to see that such things as a bowl and a reclining chair are not convex. (It is provable that a _solid_ is _convex_ if, and only if, every plane section is a convex region.)

By convention, we call the surface of a solid convex if the solid is convex. And we call a curve convex if it bounds a convex region. Rectangles and circles are convex curves, as are angles of less than π radians. (Sometimes people use the word convex to describe relative position. This popular usage of the term refers to an open curve as convex toward a certain point or line when it bulges toward that point or line.)

Your work on projections from a point confirmed that projection from a point preserves the number of sides and the convexity (or non-convexity) of a plane figure.[1] (It also happens that parallel projection modeled by the sun's shadows preserves the same two properties.)

In topology, we were not concerned about maintaining convexity nor the number of sides. Sets of points topologically equivalent need not have this much in common. For instance, a circular region is topologically equivalent to both a convex polygonal region and a nonconvex polygonal region; all three are in the same topological equivalence class. Projection defines equivalence classes of things that are topologically equivalent and that also have the same number of sides and convexity. Projective equivalence is more specific than topological equivalence. If two sets of points are projectively equivalent then they are also topologically equivalent.

EXERCISE 5:

(A Discrimination Game)
Bring to class a model of a solid or a closed surface

---

[1] Convexity is maintained because the projection of each line segment lying within the modeled region lies entirely within the region's shadow. Hence, if the original is convex, so will be its shadow.

from the world of objects. (Limit the size to approximately 1 foot in the greatest dimension.) Put all things on a table for examination and handling by the class members. Set up a slide projector near the table, facing an empty wall and far enough from the wall so that the class members can all form a group in front of the projector for a shadow show. Make the game a self test. Each class member will project a shadow of an item that he chooses from the table. The others try to guess the item he's holding by looking only at the shadow. Fill out a test form with three spaces on each line—one for the name of the individual displaying the shadow and two for your first and second guesses of the object he holds. To proceed, take turns picking out an object from behind the group and holding it before an empty projector; turn on the projector and sustain the position for ten seconds before turning off the light. The class enters their guesses by the name of the student who showed it, for he will be the authority when they start to compare answers. A suggested scoring system awards two points for having the correct answer as a first guess, and one point for having it as a second choice.

This discrimination game may lead to a discussion of shadows of 3-dimensional surfaces and solids. It is fun to use such models when first exploring projection, but this method is not as efficient for uncovering the two aforementioned properties of projection, preservation of convexity and number of sides. The problems are: 1) The preservation of convexity shows up when using 3-dimensional objects, but not the preservation of non-convexity.

(See Exercise 6) 2) The number of faces will not be indicated by the shadow because not all the faces can be illuminated.

In 2 dimensions, there is a one-to-one correspondence between points of the original and points of the shadow. When projecting a 3-dimensional set of points onto a plane, many points of the object are projected onto a point of the shadow; it is not a one-to-one correspondence. The failure of a 3-dimensional nonconvex object to have a non-convex shadow is a result of the many-to-one property of this projection, and not to some mysterious property of 3 dimensions.

Mathematicians avoid these ambiguities by defining projection in algebraic terms so that one can discuss a projection of a 3-dimensional object onto a 3-dimensional "shadow"; such definitions form the basis of formal projective geometry.

EXERCISE 6:

Find a common household object that is not convex and yet has shadows that are convex regions.

This concludes play with projection. In this chapter you have been introduced to projective geometry, the properties of which are more specific than those of topology. This is the distinction indicated pictorially by the diagram included in the preface. In this diagram, the original triangle is projectively equivalent to fewer curves than under topological equivalence.

ANSWERS TO EXERCISES

1. a) 1) <u>N</u> 2) <u>S</u> 3) <u>S</u>   4) <u>N</u>   5) <u>N</u>   6) <u>S</u>   7) <u>N</u>   8) <u>N</u>
2. a) 1) <u>S</u> 2) <u>S</u> 3) <u>N</u>   4) <u>N</u>   5) <u>N</u>   6) <u>N</u>   7) <u>N</u>   8) <u>S</u>
3. a) 1) <u>N</u> 2) <u>S</u> 3) <u>N</u>   4) <u>S</u>   5) <u>N</u>   6) <u>N</u>   7) <u>N</u>   8) <u>N</u>

1. b) Never rounded arcs. Not necessarily perpendicular. Rarely, for the projection does not preserve proportions of the sides.

2. b) Yes, except in one case we shall call the exceptional.

2. c) No, it is impossible except where the original and the shadow coincide.

3. b) No. Yes, there is no line segment in the projection of a circle (except in the exceptional case).

4. a) All shadows of a convex region are convex.

   b) No shadows of a nonconvex region are convex.

6. A mixing bowl models such a nonconvex solid.

BIBLIOGRAPHY

Ptak, Diane M. *Geometric Excursion*. Oakland County Mathematics Project. Oakland, Mich., 1970, pp. 87–92.

Smart, James R. *Introductory Geometry—An Informal Approach*. Monterey, Calif.: Brooks/Cole Publishing Co., 1969, p. 184.

Ward, Morgan, and Hargrove, Clarence Ethel. *Modern Elementary Mathematics*. Reading, Mass.: Addison-Wesley Publishing Co., Inc., 1964, pp. 181–5.

chapter 10

congruence

&

similarity

MATERIALS NEEDED:

Compass and ruler for each student

5 or 10 protractors

Candle, pin, and box

Yardstick

5 rubber balloons

Tracing paper

In this chapter, we will explore the rigid motions Euclid had in mind when he defined congruence. You will use what you learn about congruence to state minimal instructions to others and to construct things. Similarity, a somewhat more general geometry than congruence, will be explored using several visual aids.

10.1 CONGRUENCE AND INTUITION

Of the many relationships between geometric figures, that of congruence is probably of use in more areas of our civilized society than any other. The child practices congruence when fitting a piece of a jigsaw puzzle into the correct hole and graduates to the workaday world of constructing things from paper patterns and buying replacement parts when items break down. The textile to be cut out for a

FIGURE 1

dress must be congruent to the pattern on paper, and that is why you pin them together before cutting out the textile. The new fender for your auto is to fit perfectly, and that is why you buy a new one for a certain year, make, and model. Because the new fender is congruent to the discarded one, your neighbors need never know that you pulled a dumb stunt backing out of your driveway.

This putting of one thing in the place of another is what Euclid meant when he introduced the idea of geometrical congruence. Euclid's definition is intuitive, for geometry actually concerns sets of points rather than objects that can be moved.

DEFINITION

Euclid defined figure *A* as <u>congruent</u> to figure *B* when figure *A* could be moved to coincide with figure *B* so that every point matched.

A mathematical definition of <u>congruence</u> terms it a particular topological equivalence in which all distances between points are kept fixed.

Regardless of which definition you find yourself using, you should find yourself identifying as congruent those sets of points with the same shape and size. It is from these characteristics that we get the symbol for congruence, ≅. Congruence, ≅, means equal size and similar, = and ~ put together.

EXERCISE 1:

Check your awareness of congruence by trying to

guess all the sets of congruent quadrilateral regions in Figure 1. Put a numeral *1* on all those regions in one congruence class, and a numeral *2* on those in still another class, and so on for other congruence classes.

The next step after recognizing congruent figures is to sketch some. Try drawing some line segments and angles, congruent, respectively, to each of those below. Proceed, using pencil sketches, making each drawing as different from the original as you can.

Are you remembering from Chapter 9 how to move a line segment and how to move an angle? A fast and accurate way to move a line segment is to match the endpoints of the original line segment to the points of the compass and move the compass to a new location. Similarly, with folded paper or a compass you can construct, anywhere, an angle congruent to the above angle. And both the line segment and the angle can be moved, using the intuitive definition, by means of a tracing of the original. Check your drawings to be sure that you have shown that congruence has nothing to do with position.

10.2 REFLECTIONS, TRANSLATIONS, AND ROTATIONS

Although a great structure has been built up since his time, Euclid actually meant by congruence of

two figures that one can be moved (as on a transparent sheet) to coincide with another. When he used the word "move," he was thinking of certain motions (mappings). (His motions did not include those we now call projectivities or the elastic motions of topology.) After Euclid's time, mathematicians deduced the motions Euclid had in mind; they are called rigid motions. If it is assumed that distance is preserved by certain motions, then it is possible to prove that all such motions are made up of a combination of rotations, reflections, and translations. This is why these three motions are explored in many geometry books.

FIGURE 2

Reflection is one of the rigid motions. (For point sets in the plane, use of the lone word "reflection" is understood to mean reflection about a line.) You are familiar with reflection about a line from your experimentation with symmetry in Chapter 2. A figure and its reflection appear symmetrical about some line. Reflection about a line can be better defined now that the words "length" and "perpendicular distance" can be used. We say that the figure on the right in Figure 2 is a *reflection* about the vertical line of the figure on the left, and vice versa. So point for point, lines drawn from the left curve, perpendicular to the line of reflection, pass through corresponding points in the right curve, and these corresponding or dual points are equidistant from the line of reflection. Try visualizing horizontal and vertical reflections of each of the quadrilateral regions in Figure 1. If you aren't sure of these, return to the section on symmetry in Chapter 2.

Translation, another rigid motion, is the movement of one figure onto another performed by sliding the figure without reflecting or rotating it. For example, when a paper is left stable in the typewriter, a capital *L* in one place will be a translation of a capital *L* anyplace else on the paper. And, of course, any letter typed one place on the sheet is a translation of the same letter typed anywhere else on the sheet. If an *L* is seen upside down or twisted slightly, it is not a translation of the first.

EXERCISE 2:

Figure 3 is a collection of figures that appear similar to each other. Some are translations of each other, some are reflections of each other, and some are neither.

a)   Classify all figures into sets of all those you could show congruent by translation; then repeat the total classification into pairs where the second is a reflection of the first.

b)   Indicate those pairs in Figure 3 where one is obtainable as a combination of one reflection and one translation of another.

c)   Are reflection and translation generally commutative? If so, a figure should end up in exactly the same place after both combinations, reflection-translation and translation-reflection. To test this, sketch a simple figure and compare the result of reflection-and-then-translation with the result of translation-and-then-reflection when you have a certain reflection and a certain translation in mind. One test starts with reflection about a vertical line,

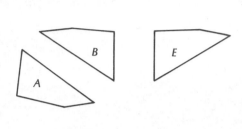

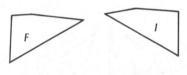
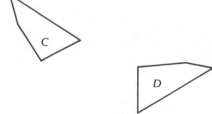

FIGURE 3

yielding A_1 in Figure 3'. Complete the reflection-translation by drawing in the 2-inch horizontal translation to the right, and by labeling it A_2. The translation-reflection is also begun for you by showing B_1. Complete the figure by drawing in B_2 and see if $A_2 = B_2$, a necessity for commutativity.

The rigid motion called rotation is perhaps the most subtle to recognize. Perceiving a figure as lying on a wheel allows you to see easily a set of all those figures into which it can be rotated as the wheel turns. If you imagine the same figure on a bigger disk—the center further away—you get another set of those figures congruent by rotation, and so forth for each circular disk you imagine your figure lying on. If one figure can be thought of as spun around on a disk until coinciding perfectly with another figure, we call one a *rotation* of the other.

Examine Figure 4 and you will see that in each picture the shaded portions are indeed rotations of each other. In this case the rotation is around a point of intersection of the regions.

Figure 5 shows four pairs of regions congruent by rotation. Note that the center of rotation can be inside, on, or outside the figure rotated.

EXERCISE 3:

See if you can devise a method for finding the point of rotation in Figure 6—a method that will apply to any figure. With a method, you can find the point and thus show congruence by rotating one

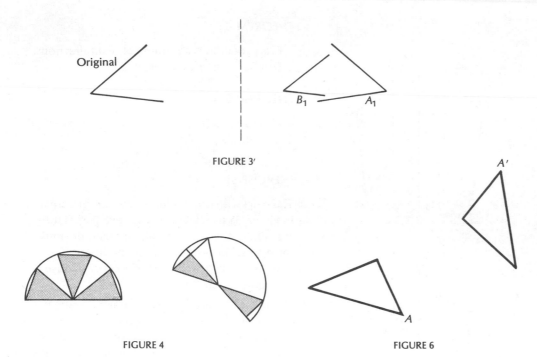

Original

FIGURE 3′

FIGURE 4

FIGURE 5

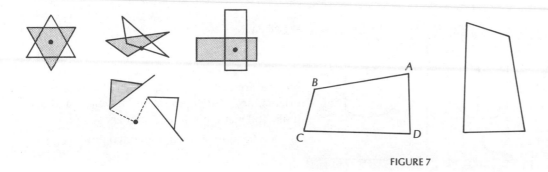

FIGURE 7

B_1 A_1

A'

A

FIGURE 6

A

B

C D

figure to a position over the other.

Can you describe a method by which you found the point of rotation in Figure 6?

If not, can you at least sketch a line on which the point of rotation must lie?

We assume (as Euclid did) that certain motions of geometric figures will preserve distances between points—thus preserve the length of the line segment and the amount of angle opening. It can be proven that all the motions assumed to preserve distance can be expressed as a combination of reflections, translations, and rotations. So, any motion by which you move one figure to show it congruent to another is capable of expression as one of these three or a sequence of them performed consecutively.

Figure 7 shows a rigid motion (a translation followed by a 90° rotation) of *ABCD*. With a compass, you can affirm that respective line segments and diagonals have equal length before and after the rigid motion pictured in Figure 7. In addition, you can review several methods of checking the opening of an angle *A* while affirming that it is the same size in the image as in the original.

EXERCISE 4:

Consider the shape of the capital letter, *L*. Could any figure congruent to this be shown congruent

a)   by a sequence of translations?

b)   by a sequence of reflections?

c)   by a sequence of rotations?

Give examples and counterexamples for each answer.

A sequence of translations is not sufficient to move a figure so that it coincides with any other that is congruent. Neither is a sequence of rotations sufficient, as you no doubt found out in Exercise 4. Also insufficient are two-move sequences such as translation-reflection and rotation-reflection, even when the order is unspecified. Turn to Exercise 26 for guidance in exploring the possibility of rotation and translation to show congruence by coincidence.

Your experience with the three rigid motions should allow you to see that any figure can be moved in Euclid's sense, by a combination of these motions, to be shown congruent to another. Exercises 26–29 give practice in this. You have seen, as a result of working on Exercise 4, some rigid motions that will work all the time. You should be prepared to accept as true the following theorems:

THEOREM 1

Every rigid motion is a product of rotation, translation, and reflection motions.

THEOREM 2

Every translation is a product of two reflections in parallel lines.

THEOREM 3

As a result of Theorems 1 and 2, every rigid motion is a product of rotations and reflections.

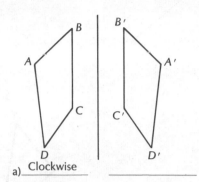

a) Clockwise

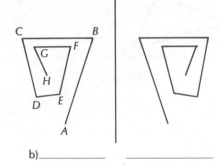

b)

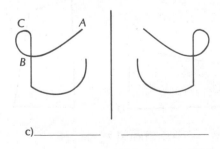

c)

FIGURE 8

THEOREM 4

Every rotation is a product of two reflections. (This is the hardest to believe.)

THEOREM 5

As a result of Theorems 1, 2, and 4, every rigid motion is a product of reflections.

EXERCISE 5:

The clockwise orientation of the left quadrilateral in Figure 8a (clockwise as your eye passes through the points A, B, C, D, in that order) is obviously maintained in a translation of the figure. We say that the original and its translation each have clockwise orientation.

a) In each figure below, designate, by the word "clockwise" or the word "counterclockwise," the orientation of the original (consider it to be the one to the left) and the orientation of the reflection map lying on its right.

b) Does reflection about a line always reverse the orientation of a plane figure?

EXERCISE 6:

Consider the labeled "square-spiral" on the left of Figure 8b. Below, mappings of this figure are described; for each mapping, state "always preserves orientation," "sometimes," or "never"—and back it up. The mappings to consider are

a) a reflection about a horizontal line,

b) a reflection about the line through *B* and *H* in the spiral,

c) a rotation clockwise, and

d) a rotation counterclockwise.

In three dimensions, translation and rotation are movements obviously analogous to their counterparts in the plane. Any figure can be rotated or translated in space, and the mapping found to be congruent to the original and of the same orientation as the original. Try waving about a shoe box with a shoe in it, and you will see that rotation and translation leave intact both the shoe box and the position of the shoe with the heel next to a certain corner. Reflection in 3-space takes place through a plane, just as in mirror symmetry. A shoe box is its own reflection across a plane extending down its middle; a shoe box on one end of your desk is a reflection of an identical shoe box at the other end through a plane slicing through the middle of your desk.

EXERCISE 7:

Would a pebble lying in a corner of the shoe box on your left be a pebble in the same corner in its reflection across the plane cutting through the middle of your desk?

Will the heart on your left side be on the right for the person who is your true reflection?

Discuss orientation of the reflected object compared to the original.

Do you perceive this mapping, analogous to mov-

ing a set of points, as continuous or discontinuous?

Because of the peculiar effects reflection has on orientation in two and three dimensions, some theorems in mathematics exclude reflection directly or indirectly. Here's one such theorem, believeable although limited in scope:

THEOREM

Every orientation-preserving rigid motion can be expressed as a product of one rotation and one translation in any order.

10.3 STATING INSTRUCTIONS

The traditional geometry course approaches the study of congruence by more or less stating for you what is known, and then encouraging you to use that information whenever you want to prove two figures congruent. So, with triangles, the traditional initiation includes statements (theorems) describing in detail what it takes to prove two triangles congruent. An equally fruitful and stimulating approach is to ask what minimum amount of information would be needed to construct a triangle, unique except for position. There is a connection between proving two triangles congruent and giving instructions for construction. We shall study congruence by thinking through what minimum set of instructions would specify a particular triangle (and all those congruent to it). For instance, would a particular triangle *ABC* be indicated by giving three instructions: the length of $\overline{AB}$, the length of $\overline{AC}$, and the measure of angle B? Will these three instructions suffice or might they lead to an impossible situation in which no triangle can be constructed or an ambiguous case where two or more triangles are indicated? To check this, put down some line segment on the horizontal and call it $\overline{AB}$; then continue the construction from there, correctly labelling the three vertices.

EXERCISE 8:

Take time to draw a triangle or a collection of them that satisfies the written directions above, if $m(\overline{AB}) = 4$ inches, $m(\overline{AC}) = 3$ inches, and angle B is one-sixth of a straight angle.

For construction of a particular triangle, in no matter what position, we can develop a minimal set of instructions. Obviously, by writing the multitude of sentences necessary to specify the measures of all the line segments and all the angles, we can make sure someone constructs the same triangle, but it is more sporting to do this with three sentences about measure—or four if necessary. You found out in Exercise 8 that the three written directions given were not sufficient to determine a particular triangle; you ended up with two noncongruent triangles that fulfilled those conditions. The next exercise involves giving you a verbal description and asking you to draw, whenever possible, a particular triangle to meet that description or a series of triangles that do. Do this for each verbal description below.

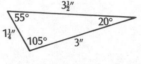

FIGURE 9

EXERCISE 9:

Are each of the sets of instructions given below sufficient for you to construct a particular triangle? If not, construct all noncongruent triangles for which that set of directions is true, or indicate verbally the set of triangles, after drawing a few examples.

Set 1: $m(\overline{AB}) = 4''$
  $m(\overline{BC}) = 2''$
  $m(\angle A) = \frac{1}{3}$ of a straight angle.

Set 2: $m(\overline{AB}) = 4''$
  $m(\angle B) = \frac{1}{6}$ of a straight angle.
  $m(\angle C) = \frac{2}{3}$ of a straight angle (120°).

Set 3: $m(\angle A) = 45°$
  $m(\angle B) = 45°$
  $m(\angle C) = $ a right angle.

Set 4: $m(\overline{AB}) = 4$ cm.
  $m(\overline{BC}) = 3$ cm.
  $m(\overline{AC}) = 2$ cm.

EXERCISE 10:

Now you try your hand at making up instructions. Note Figure 9, and make sets of instructions that will guarantee that the reader draw the same triangle (congruent to Figure 9). Limit each set of instructions to three stipulations—three simple sentences about measure like those in the sets of Exercise 9. Give as many sets of instructions as you can.

The minimum information for construction of a triangle is identical to the minimum that one would have to check in comparing two triangles for congruency.

The connection between stated instructions and conditions of congruence may have dawned on you when working on Exercise 10. If you felt sure of a set of 3 instructions, then you would only check over those 3 measurements when looking to see if the reader drew the correct triangle. Man has made for posterity the shortest possible statements about congruent triangles—theorems such as, "Two triangles are congruent if, and only if, the measures of two sides and the included angle are equal to the measures of the corresponding parts in the other triangle." One can consider these theorems as checks for congruency or as minimum requisites in constructing one triangle congruent to another.

Similar brevities in instructions for construction exist for other polygons. In Exercises 30 and 31 at the end of the chapter, you will have a chance to develop the briefest possible instructions for congruent parallelograms and rectangles.

10.4 CONSTRUCTIONS BASED ON CONGRUENCY

In the last section we made up short-cuts, theorems based on Euclid's notion of congruency of two figures. In this section, we will use the axioms of Euclidean geometry to construct things.

The bisections of an angle and of a line segment are constructions based on common sense notions of congruency. Most likely, you can, by yourself, bisect ∠A in Figure 11, that is, find a line through the middle of the angle. If not, study Figure 10.

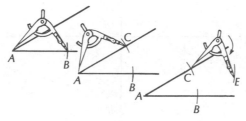

FIGURE 10

The Beginning Movements in the Construction of an Angle's Bisector

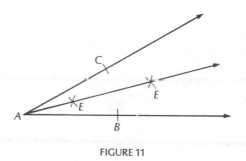

FIGURE 11

When somebody says "bisect the angle," this describes a whole sequence of permissible constructions—permissible by the axioms. The axioms of Euclidean geometry give us the right to

1) construct a line through two points
2) transfer a line segment
3) draw a circle with a given center and radius
4) find a point of intersection when existent, of
   two lines
   line and circle
   two circles.

Combinations of these comprise the so-called Euclidean constructions. Our yield might be no learning at all if we approached this topic by drawing an angle and saying "bisect it." Try this approach instead: Study Figure 11 for a few minutes. Try to reconstruct the steps in that construction. See if you can use previous knowledge of congruent figures to convince yourself that $\overrightarrow{AE}$ really bisects the original angle. Should this marking up of Figure 11 leave only hazy knowledge, read on for more help. In case you do not yet see your way clear to duplicate the bisection in Figure 11, I will describe the traditional construction step by step and ask you in Exercise 11 to try to explain just why this construction proves $\overrightarrow{AE}$ bisects angle A.

To construct the bisector of ∠A, the given angle is marked with a compass at points B and C, which are the same distance from A on the two rays. Then from B and C, equal arcs are made out in space, crossing in point E (or E'). Draw the ray from A through E (or E'), and it will bisect ∠A.

EXERCISE 11:

Examining the completed angle bisection in Figure 11, explain why $\overrightarrow{AE}$ bisects $\angle A$. This necessitates specifying those line segments and angles that are congruent.

A line segment is bisected by a line cutting through its center. The bisection of $\overline{DE}$ is effected by making the line $\overleftrightarrow{AB}$. A compass is opened with the point at D—opened enough to reach generously past the midpoint of the segment $\overline{DE}$—and arcs are made above and below the approximate midpoint. The compass point is then placed at E—without changing the opening—and arcs are marked above and below so that they intersect the previous pair in points A and B. In Figure 12, line $\overleftrightarrow{AB}$ bisects line segment $\overline{DE}$ at the point M. M is called the midpoint of $\overline{DE}$.

EXERCISE 12:

a)   Draw four additional lines in Figure 12 to obtain Figure 13.

b)   Refer to congruent triangles to explain why $\overleftrightarrow{AB}$ bisects $\overline{DE}$. (Be careful that you make no statements merely on conjecture or faded memories.)

c)   Can you assert any sure characteristics of the

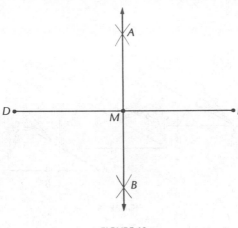

FIGURE 12

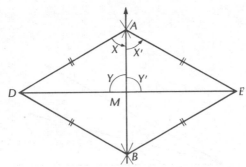

FIGURE 13

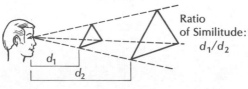

FIGURE 14

four angles meeting at the midpoint, M? If so, specify.

d)   Would $\overleftrightarrow{AB}$ bisect $\overline{DE}$ if the compass opening were changed when moving the point to E?

By leaving your compass opening the same in marking the four arcs, you have constructed four congruent line segments—$\overline{DA}$, $\overline{DB}$, $\overline{EA}$, and $\overline{EB}$. Having realized that $\overline{AB}$ is congruent to itself, you can observe that the left triangle, ADB, is congruent to the right one. Consequently, $\angle X$ is congruent to $\angle X'$ because they correspond in reflected triangles. Since this is so, the two triangles above $\overline{DE}$ are congruent. Why so? Thus $m(\overline{DM}) = m(\overline{ME})$ because these segments are matching sides of two congruent triangles. Hence, point M is the bisector of $\overline{DE}$.

This construction yields a bonus—that the bisector has also been constructed perpendicular to $\overline{DE}$. The angles Y and Y′ formed above $\overline{DE}$ are of equal measure because they, too, are matching parts of the same congruent triangles. Hence, they are each half of a straight angle, i.e., they are right angles. This is why we know that $\overleftrightarrow{AB} \perp \overline{DE}$.

10.5   SIMILARITY

For pedagogical reasons, similarity is developed after congruence because similarity demands greater perception of subtleties and because we can then use what we know about congruence, parallelism, and perpendicularity to develop the properties of similarity. Similarity is more general; congruence will be seen as a particular case of similarity.

We can use the idea of projection from a point (the student's eye) to develop the notion of similar figures. Distances are measured, as indicated in Figure 14 to find a ratio of similitude.

A way to approach similarity—a way that will yield concrete information—is to work with matching. If you take a quarter and a dime, kneel down at the edge of your desk, and move the two coins on a line with your eye, you will find it is possible to line the two up so that the edges just *match*. (Re-examine Figure 14). Is it possible to get the coins to *match* if you place the quarter toward you and the dime away from you?

EXERCISE 13:

a)   Try matching a few plane objects such as post-cards and covers of books. List each pair of plane objects compared, and report yes or no on their "matchability".

b)   When you are holding one matching pair, ask a friend to measure the distances d_1 and d_2, as shown in Figure 14. You will find the ratio of these two numbers identical to the ratio of any of their dimensions, length of diagonal, length of side, what-ever. Record d_1, d_2, and a pair of corresponding dimensions and check the theory that the ratios are the same.

Two plane figures are *similar* if they can be matched when held parallel. Note, in reexamining Figure 14, that respective points on similar figures lie on a line from the eye. Call the eye point P and visua-

FIGURE 15

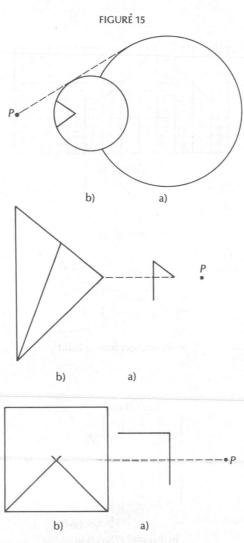

lize yourself putting the smaller of the similar figures in place so that its respective vertices etc. lie on lines from P to the larger figure. We will use point P to aid you in the following exercises as you ascertain similarity and learn to construct one figure similar to another.

EXERCISE 14:

Would it be possible for a polygon of four sides to be similar to a triangle?

Can an isosceles triangle be similar to a triangle in which all three sides are of different length? (An *isosceles triangle* is any triangle having two sides congruent.)

EXERCISE 15:

a)   Match figures a and b in each of the three pictures in Figure 15, and complete figure a of each so that it is similar to the corresponding b. You may be able to imagine your eye at point P. To match physically, the smaller figure should be on a transparency so you can tell when it matches the figure held behind the transparency. No matter what pair you hold up, you will see where two figures match well enough to sketch, with a felt pen, the remaining lines of figure a.

b)   Use your visualization of the matching process to complete the small doghouse started in Figure 16, purposely drawn on graph paper to aid your visualization.

As you probably noticed from this last exercise,

photographic enlarging and reducing is one application of the concept of similarity. Proportions must be maintained in an enlargement; goodness knows, we would all quit asking for enlargements if they showed our loved one's arms or face stretched out of proportion to the rest of the picture. If you ask for a one-to-three reduction of a print so that it will fit in your billfold, you are asking that every 3 units in the original become one in the reduction. We request the ratio four-to-three (4:3) when we want the photographer to give us a print 4 units high instead of 3 and 4 units wide for every 3 units of width it had previously.

EXERCISE 16:

a)   Does the arithmetic ratio, 4:3, describe any of the similarities shown in Figure 15? or Figure 16? If so, sketch a pair with that ratio and label the lengths on each of the line segments in the sketch.

b)   Do all respective line segments of such a pair, regardless of their slant, have the ratio 4:3? Explain by comparison to the photo enlargement of height and width.

c)   Can you guess, before experimenting, the comparative distance from the eye to the smaller of two figures whose line segments were in the ratio of 4:3? How does the distance from eye to smaller compare to the distance from eye to larger? You may want to look back at the table in Exercise 13 and reexamine the figures to see whether there is any correlation between distance from the eye and lengths of segments. Check your conjecture or write up an explanation as to why your guess must be correct.

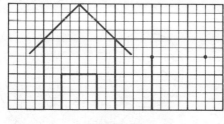

FIGURE 16

FIGURE 17

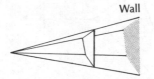

Wall

a) Projection from a Point

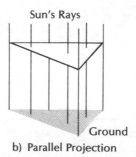

Sun's Rays

Ground

b) Parallel Projection

In Chapter 9, we played with shadow projections of polygonal regions. We found that no matter where we held the polygonal region with respect to the light source, with the exception of one position, our shadows were not only topologically equivalent, but also maintained the number of sides and convexity of the original. Figures as alike as a projection and its original have a lot in common, yet they are not called similar. Mathematicians reserve the word _similar_ to define a more specific relationship. It appears logical that a limitation on the movement of the patch before the screen might demonstrate similarity. It does! If we limit our movement of the polygonal region so that it is always parallel to the wall, then no matter where the light source, the shadow is similar to the original. Parallel movement of the original, forewards and backwards, yields different similar shadows.

Our shadow more nearly approaches the size of the original when we hold the region near the wall. Can our shadow ever be the same size?

In projection from a point, the shadow shrinks to the same size at the instant the region touches the wall. However, in working on ways that the shadow and the original could be the same size, you may have discovered another source of light, sunlight. Sun's rays appear essentially parallel because the sun is so vast a light source. Thus they approximate a different kind of projection called parallel projection. (See Figure 17b.)

Projection from a point is our present mathematical

topic. Besides, parallel projection, yielding nothing but congruent shadows, is so confining and boring a study that we dismiss it summarily. Henceforth, we shall use the word _projection_ to mean projection from or through a point.

EXERCISE 17:

Can you describe any way of getting a point projection to be twice as big as your patch? Half as big? The same size? (Experiment with different placements of the point relative to the two patches.)

A light source projecting a shadow and an eye lining up two similar shapes that match when held parallel—these are both physical analogs of projection from an _outside_ point, shown in Figure 17a. However, more freedom to explore is possible when abandoning the idea of light or eye in conceiving of point projection. If you merely draw lines from the original through a point until they hit a parallel plane, is there any way you can get a projection smaller than the original? Is there any point through which you can project and get an upside-down image? Yes, each of these can be accomplished by choosing a point between the two figures. See Figure 18 for an example of projection through an _inside_ point. Physical demonstrations of projection through an _inside_ point are limited; however, because this demonstration yields such an unexpected result, it is worthwhile having one class member prepare a demonstration of a candle in a box showing an inverted flame to the eye seeing it through a pinhole.

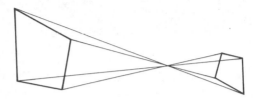

FIGURE 18

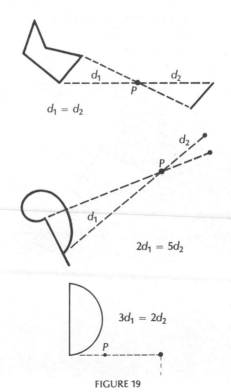

FIGURE 19

Since similarity is also maintained by projection through a point inside, the definition can be reworded. Plane figure _A_ is <u>similar</u> to plane figure _B_ if the planes containing _A_ and _B_ are parallel and if there is a projection through a point that maps points of _A_ onto points of _B_.

EXERCISE 18:

a)   Given a curve and a point of projection in the plane, complete the sketch of the projection of each part of Figure 19.

b)   Sketch a point of projection by which the semi-circular curve in Figure 19 is shown to be congruent to its projection or explain why that is impossible. Can you generalize?

c)   There is a certain comparison of sizes of similar figures when d_1 is less than d_2. What is true of the figures when $d_1 = d_2$?

d)   Is a projection of a circular patch ever anything other than circular? Explain in detail.

Draw a triangle on a piece of rubber balloon and trace it on a piece of paper for later reference. Topological equivalence is maintained by any stretch of the balloon. A limitation on the stretching will produce a figure similar to the original. Experiment with stretchings of the triangle drawn on thin rubber to find a stretch that yields a similar figure.

EXERCISE 19:

What stretches of the rubber seem to yield a triangle

similar to the original? Experiment! Compare the result of a vertical stretch with the traced original, and sketch the two side by side. Experiment with two-way stretches where the directions of stretch are perpendicular to each other to see if you can come up with anything similar to the original, and draw the two side by side. Draw or describe a stretch that yields a similar figure. (Drawings tend to be rather exotic.)

Limitation of a topological transformation—specifying the stretch—is a way to define similarity. The two way stretches that stretch an equal distance in two perpendicular directions preserve similarity; they are called _conformal transformations._ The photo miniaturizing in a ratio of 3:1 is an example of a _conformal transformation_, that is, a shrinking along both horizontal and vertical axes.

We have seen figures that were similar and were affirmed so by placing them in certain positions to show a point projection, a shadow, or a conformal transformation. We leaned on intuition and placement in space. If we restrict attention to polygons, we might discover characteristics in common between two similar polygons, regardless of their relative positions in space. There must be some connection such as the sizes of angles or the lengths of corresponding line segments. Figure 20a shows a pair of similar triangles, obviously similar by point projection from A. The pentagons in Figure 20b are also similar for the same reason. Seeing similar figures _nested_ in this way and asking ourselves some questions may help us state a set of conditions for similarity of two polygons.

FIGURE 20

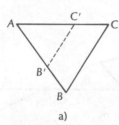

a)

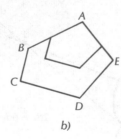

b)

EXERCISE 20:

a) Can the similar triangles in Figure 20 be repositioned so that they are coincident at an angle other than angle A? Show this or explain why they can not.

b) Are the respective angles of similar figures congruent?

EXERCISE 21:

a) Is the congruency of respective angles sufficient to guarantee similarity? For two triangles? For two quadrilaterals?

b) Show that the congruency of four angles of a quadrilateral to the respective angles in another quadrilateral does not imply similarity.

DEFINITION OF SIMILAR POLYGONS

Two polygons are similar if their vertices can be paired so that

1) the measures of corresponding angles are equal, and
2) the measures of corresponding sides are proportional.

You have applied the above definition even if you haven't articulated it. Looking back over your perceptions when exploring projections and matching with transparencies, you may remember noticing some things: You saw that angles were congruent when you lined up two figures with your eye and pronounced them similar, and you may have checked the accuracy of your drawing by checking to see if the line segments were proportionally larger or smaller than those in the original. No doubt the proportionality of the length of line segments enlarged in Figure 16 was the basis on which you were able to go ahead and complete the picture. Angle measure and proportional sides are so interrelated that, with triangles, similarity can be proven by using two pairs of congruent angles. Experimentation with triangles leads one to other minimal sets of information guaranteeing similarity; either one angle congruent and proportional adjacent sides or three proportional sides will prove two triangles similar. Other types of polygons have their minimal conditions by which two can be proved similar. Further testing for possible theorems guaranteeing similarity of two quadrilaterals or two pentagons is fun and elucidating.

The definition above is not as minimal a definition nor as general a definition as you may wish to know. The definition below is both minimal and general; you may use it if it is meaningful to you.

DEFINITION OF A SIMILARITY MAP

A _similarity_ is a point transformation of the unextended plane onto itself that carries each pair of points (A,B) into a pair (A',B') such that $m(\overline{A'B'}) = k \cdot m(\overline{AB})$, where k is a fixed positive number.

Similarity is of great practical importance because it is the basis of much indirect measurement. In the

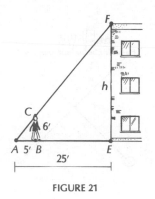

FIGURE 21

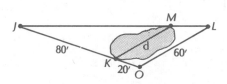

FIGURE 22

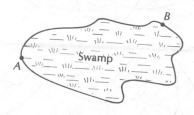

FIGURE 23

following examples, the height of a building and the distance across a pond can be computed easily.

EXAMPLE:

If we assume that the man and the edge of the building are parallel, the picture in Figure 21 suggests a conformal transformation, an expansion from point A. Triangles ABC and AEF are therefore similar.

So $5/25 = 6/h$

Thus $5h = 150$,

and $h = 30$

ANOTHER EXAMPLE:

Use the diagram in Figure 22 to find the distance across the pond. By the similarity of the two triangles passing through J,

$m(\overline{KM})/60' = 80'/100'$

$m(\overline{KM}) = 48'$

EXERCISE 22:

Explain how you could find the distance from A to B without mushing your way through the swamp pictured in Figure 23. (Hint: construct similar or congruent triangles.)

EXERCISE 23:

a) Why would we know that the TV tray in Figure 24 is parallel to the floor if we measured and found out that the legs bisect each other.

b) What if the legs are unequal in length but still bisect each other?

EXERCISE 24:

Does the set of all similar triangles form an equivalence class?

EXERCISE 25:

Are all right triangles similar?

10.6 EXERCISES FOR CHAPTER 10

26. Can you show any two congruent figures to be so by performing a translation, followed by a rotation?

   In the reverse order?

   Can you show it if order is unspecified?

   Be specific enough to at least give opinions (yes, no, or maybe) and examples to back up your opinion.

27. a) In Figure 25, determine which of the four figures on the right and below the first one on the left can be obtained from the left figure by a sequence of translations, rotations, and reflections. State how.

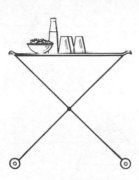

FIGURE 24

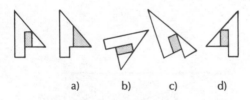

a)   b)   c)   d)

FIGURE 25

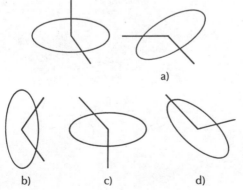

a)

b)   c)   d)

FIGURE 26

b) In Figure 26, determine which of the figures on the right can be obtained from the figure on the left by a sequence of translations, rotations, and reflections. State how.

28. An aerial view of a living room and adjacent bedroom is pictured in Figure 27. There are also some plane figures pictured in the living room. Can you remove all the figures from the room without their leaving the plane of the paper? Give directions using translations, rotations, and reflections to get each figure through the door, or state that it is impossible.

29. A curve or region point–symmetrical to another through a point in the plane is what we call an *inversion* of the original. We also say one is the *reflection about a point* of the other. Review pairs of curves symmetrical about a point, checking to see if the inversion image is congruent to the original.

   a) Does the inversion in the plane preserve all distances, and thus make an image congruent to the original?

   b) Is there any relationship between 180° rotation and inversion?

30. In each case, construct a quadrilateral or state that it is impossible. If it is possible, draw as many as you can, unless more than four are possible. In a case where more than four quadrilaterals could be drawn, draw four as different as you can and indicate the situation verbally.

a) Construct a quadrilateral with sides of lengths 3 inches, $\frac{1}{2}$ inch, $\frac{3}{2}$ inch, and $\frac{1}{2}$ inch, as you see them in clockwise order.

b) Construct a quadrilateral having one pair of parallel sides.

c) Construct a quadrilateral having two pairs of parallel sides, with one pair of opposite sides measuring 2 inches.

d) Construct a quadrilateral having two pairs of parallel sides, with all sides 1 inch in length.

31. a) In the spirit of Exercise 10, devise a minimal set of instructions for constructing a certain parallelogram. Is there another set? (Look at Figure 28, which classifies quadrilaterals.)

b) Do the same for a rectangle.

c) If you start with the directions for a parallelogram and make alterations, what set of directions would you give for constructing a rectangle?

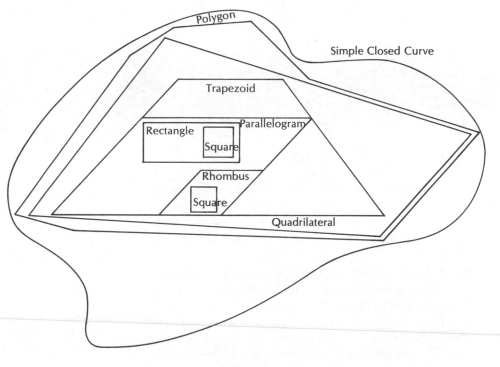

FIGURE 28

A Figure Showing the Classification "Simple Closed Curve" and Its Subsets Relating to Four-Sided Figures.
Each Term Refers to a Special Subclassification of the Next-Largest Figure (Next More-General Figure).

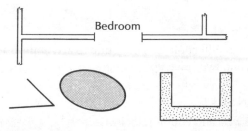

FIGURE 27

ANSWERS TO EXERCISES

1.

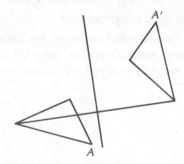

2. a) Translations: E, D, H, and F
                      B, G, and I.
   Reflections: C and I   These pairs are reflections of
                     G and H   each other about a line
                     B and E.   bisecting the line segments
                                between the corresponding
                                vertices.

   b) Naturally, all sets mentioned in answer 2 a) could be
   thought of as combinations of reflections and translations.
   Those congruent by translation could be thought of as a
   combination of an identity reflection and a translation;
   those which are reflections of each other can be thought
   of as a combination of a reflection and an identity translation (an identity translation is a translation nowhere). But
   an interesting pair is found in B and D, where B reflected
   to the right and then translated down to the position of
   D. Similar pairings with D are (G,D), (I,D), and (A,D).
   Because of the relationship of E and H with D, four like
   pairings will be possible with E and H, respectively. Continue on, if you can find more pairs where one is a product
   of a single reflection and a single translation of the other
   figure.

c) Reflection and translation are not commutative mappings. Examine this drawing and note the difference between A_2 and B_2, the final results of the two mappings in
succession.

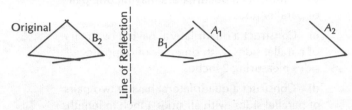

3. As a hint, I will sketch a line on which the point of rotation (center of a circle) must lie. The point of rotation lies
at an intersection of three lines; perhaps that will help
you to finish the problem.

4. a) No; Note L ⅂.
    b) Yes, two reflections can be found to accomplish any
    rotation, and two reflections in parallel lines are the same
    thing as a translation.
    c) No: Note L L.

5. a) clockwise and counterclockwise;
        counterclockwise and clockwise;
        clockwise and counterclockwise.

b) Yes, if the figure has any orientation.

6. a) never b) never. c) always. d) always.

7. The pebble changes places in the box. Likewise, a person's insides get reversed by reflection.
   There is no continuous mapping that accomplishes reflection in three dimensions. It is a point by point movement that disassembles the points and reassembles them on the other side of the plane.

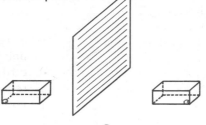

8. In miniature:

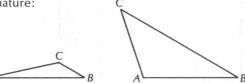

9. Set 1. Insufficient and no triangle is formed.

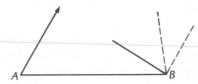

Set 3. Set of similar triangles.

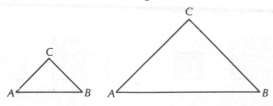

11. $\overline{BE} \cong \overline{CE}$ and $\overline{AE} \cong \overline{AE}$ and $\overline{AB} \cong \overline{AC}$.

   Hence, $\triangle ACE \cong \triangle ABE$.

   So $\angle BAE \cong \angle CAE$.

12. b) Work toward getting $\triangle DMA \cong \triangle EMA$. Then $\overline{DM} \cong \overline{ME}$.

   c) By congruent triangles, $\angle Y \cong \angle Y'$.
   Because of opposite angles, the angles below $\overline{DE}$ are congruent to those above, and hence congruent to each other.

13. a) A dime and a quarter will match; the matchbook and postcard will not.

14. No; No.

15. a) Here are two pictures completed:

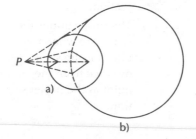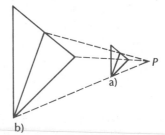

b) Small dog house completed:

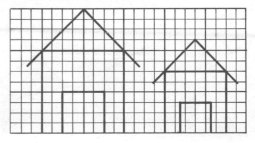

16. Figure 16 shows a ratio of 4:3 between the big house and the little. To get lengths of the respective segments in the two, examine the drawing in the answer to Exercise 15.

b) Yes, even diagonal line segments.

c) The distance to the smaller is $\frac{3}{4}$ the distance to the larger.

17. Projection from a point is twice as big if the point is twice as far away from the projection as from the object. Half as big if the point is between the object and the projection.

The projection is the same size when the object is the same distance from the point as the projection.

Half as Big

Point

Object

18. a) for $d_1 = d_2$

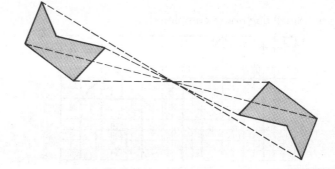

b) The only way I can get a point projection of the semicircle to be congruent to the drawing is to consider the projection plane coincident with the drawing or lying to the right the same distance.

c) When $d_1 = d_2$, then the projection is congruent to the original.

d) No. Note that the lines drawn from the point of projection map all diameters onto intersecting line segments of equal length.

19.

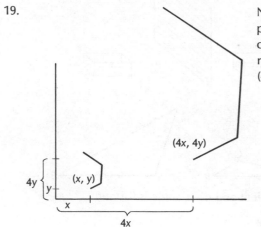

Note that each point, (x, y) is carried into a corresponding point, $(4x, 4y)$.

$(4x, 4y)$

$4y$ y (x, y)

x

$4x$

20. a) Yes, they can be placed so that another angle is seen congruent when glancing at the two figures.

b) Yes.

21. a) Not generally; however, this is sufficient for triangles.

b)

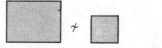 

22. Construct similar triangles having a common vertex at some point C outside the swamp. Mark points (D and E) three-fourths of the way from C to each of A and B, and then measure $\overline{DE}$. Because

$m(\overline{DE}) = 3/4\ m(\overline{AB})$,
$m(\overline{AB}) = 4/3\ m(\overline{DE})$.

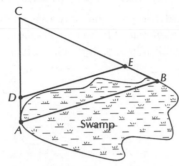

23. If we know that the legs bisect each other, we have congruent triangles formed at left and right. Hence the distance from center of wheel to tray is the same left and right. If wheels are the same size, the tray is parallel to the floor.

24. Let ~ mean "is similar to."
Reflexive property holds: $a \sim a$
Symmetric property holds: If $a \sim b$, then $b \sim a$.
Transitive property holds also: If $a \sim b$ and $b \sim c$, then $a \sim c$.
Similarity is an equivalence relation. A quick check will affirm that congruency is also.

25. No. Consider $\triangle ACB$, where $\angle C$ is the right angle. Note that $\triangle ACD$, where D is some point between C and B, is a right triangle which is not similar to $\triangle ACB$.

26. No; No; Not even that suffices for this picture.

27. a) Figure b can be obtained from the original by a translation right one inch followed by a rotation.
Figure d can be obtained by a reflection.
The others are not congruent.

b) Figure a is obtained by a translation followed by a 135° clockwise rotation.
Figure c is obtained by a translation followed by a 180° rotation.
The others are not congruent.

28. You are doing fine if you find yourself giving verbal instructions that use the proper terminology of rotation, translation, and reflection. Sliding cardboard pieces through a doorway while describing audibly will suffice, and similarly, moving a transparency over a doorway while describing audibly will suffice.

29. a) Yes. b) They are the same.

30. a) Impossible because $3'' > 2\frac{1}{2}''$, the sum of the other 3 lengths.

b) There are an infinite number.

d) Rhombus and special rhombus called a square. There are an infinite number of these rhombuses.

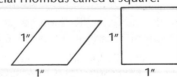

31. a) $m(\overline{AB}) = m(\overline{CD})$ and $m(\overline{AC}) = m(\overline{BD})$
or $m(\overline{AB}) = m(\overline{CD})$ and $\overline{AB} \parallel \overline{CD}$
or $\overline{AB} \parallel \overline{CD}$ and $\overline{AC} \parallel \overline{BD}$
c) A parallelogram with one right angle.

BIBLIOGRAPHY

Dotson, W. G., Jr. "On the Shape of Plane Curves." *The Mathematics Teacher* (February 1969), pp. 91–94.

Garstens, Helen L., and Jackson, Stanley B. *Mathematics for Elementary School Teachers.* New York: The Macmillan Co., 1967, pp. 145–72, 186–91.

Harting, Maurice L., and Walch, Ray. *Geometry for Elementary Teachers.* Glenview, Ill., Scott-Foresman & Co., 1970, pp. 25–26, 38–39, 43–79.

Horne, Sylvia. *Patterns and Puzzles in Mathematics.* Pasadena, Calif.: Franklin Publications, Inc., 1968, pp. 36–43.

Nuffield Mathematics Project. *Shape and Size 3.* New York: John Wiley & Sons, Inc., 1968, pp. 24–26, 48–50.

Patterson, William, Jr. "A Device for Indirect Measurements: An Entertaining Individual Project." *The Arithmetic Teacher* (February 1973), pp. 124–7.

Rosskopf, Myron F., Levine, Joan L., and Vogeli, Bruce R. *Geometry: A Perspective View.* New York: McGraw-Hill, Inc., 1969, pp. 240–53.

Schaaf, William L. *Basic Concepts of Elementary Mathematics.* New York: John Wiley & Sons, Inc., 1966, pp. 327–32.

Schloff, Charles E. "Rollings Tetrahedrons." *The Arithmetic Teacher* (December 1972), pp. 657–9.

Smart, James R. *Introductory Geometry, An Informal Approach.* Monterey, Calif.: Brooks/Cole Publishing Co., 1969, pp. 86–92.

Stubblefield, Beauregard. *An Intuitive Approach to Elementary Geometry.* Monterey, Calif.: Brooks/Cole Publishing Co., 1969, Ch. 6–8.

Usiskin, Zalman. "A New Approach to the Teaching of Constructions." *The Mathematics Teacher* (December 1968), pp. 749–57.

Whitehead, Alfred North. *Introduction to Mathematics.* New York: Oxford University Press, 1958, pp. 128–35.

Ward, Morgan, and Hargrove, Clarence Ethel. *Modern Elementary Mathematics.* Reading, Mass.: Addison-Wesley Publishing Co., Inc., 1964, pp. 185–9.

chapter 11

area

MATERIALS NEEDED:

Rulers

Postal scale and heavy cardboard

Tracing paper

Compasses

Scissors

When we measure the length of a curve, we determine how many units of some standard linear measure are contained in it. Similarly, measuring the area of a bounded plane region is a matter of determining how many of some standard unit the region contains. As curves are measured by standard linear measure—some length of a line segment, so surfaces are measured by a standard unit of area measure. The unit of measure selected should, of course, be small enough so that it will fit into the region being measured. The area is then equal to the number of times that the measuring unit is used in order to "cover" the surface completely. In this chapter, we will concern ourselves with finding the area of a bounded plane region, in particular, the area of circles and polygons. The Pythagorean theorem is also discussed in this chapter because its proofs are generally based on area.

11.1 THE UNIT OF MEASURE

Desire to cover the surface limits our consideration of possible measuring units to certain regions. Such

Note to Prospective Teacher:

Children's activities include similar efforts, counting to find area of tiled floors and ceilings and approximating area using erasers, sheets of newspaper and cancelled postage stamps on surfaces of appropriate size. Point out a surface and let children suggest which measuring unit they should use, e.g., stamps for small surfaces such as bookcovers, and newspaper sheets for large surfaces such as corridors and classroom floors. Provide a sufficient number of each measuring unit that the child can physically check his estimate without having to make pencil marks.

things as circular regions and bottoms of feet are ruled out for they are not even usable in making tessellations. (Remember Chapter 4?) Amongst those usable in tessellations are some regions extremely difficult to draw or fit together—such as the man on horseback.

EXERCISE 1:

A wall might be measured as $4\frac{1}{4}$ bookcases or $7\frac{1}{2}$ ecology posters or 500 men on horseback. These measurements are made by finding how many fit in a tessellation pattern to fill up the wall. Try it on smaller regions, filling in the table below.

Measuring unit	Surface measured	Guess	Approximation to the nearest quarter unit
Sheet of paper	Desk top		
Footlong ruler	Sheet of paper		
Quadrilateral	Sheet of paper		

a) Guess the area of your desktop at school as so many sheets of notebook paper; then measure the desktop at school, approximating to the nearest quarter sheet.

b) Repeat part a), measuring a sheet of notebook paper first, with the flat of a foot long ruler and then the quadrilateral pictured here. ⊕

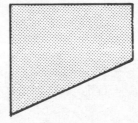

All the regions used for measuring in the previous exercise form tessellations, but not all of them would be equally good as a standard unit of measurement. That shape we want (for a standard unit of measure of area) should require no genius to fit together (which rules out oddballs like irregular polygons and men on horseback), should be easy to count (not as long and narrow as the ruler), and should be easy to enlarge without changing its shape. The square is particularly easy to enlarge to another square—for instance, one that contains 4, 9 or 16 times the area. Also the square has a special shape so that it is unnecessary to turn the square patches to estimate fractional parts. Part of the ease of fitting square regions together is that they look the same from different positions and have symmetry and "nice" dimensions.

EXERCISE 2:

a) Note the lines of symmetry of the square by folding a square patch so that corresponding points will be folded together. Verbally or pictorially report the lines of symmetry of the region.

b) Repeat the above for two different rectangles that are not squares.

c) Repeat the above for an equilateral triangle, the most symmetric of all triangles.

EXERCISE 3:

In one physical movement with a sheet of notebook paper, locate the four vertices of the largest square region you can make out of that sheet. Report what you did and why it works.

For all these reasons, and more, the square region has been chosen by earthmen as the standard unit

with which to measure the area of a surface. How large a square each man uses depends upon the size of the surface being measured, the need for accuracy, and the customs of his nation.

EXERCISE 4:

A particular geometry workbook[1] phrases its request for the area of a region by referring to a region such as that in Figure 1 and asking the following type of question: How many times is the dark square needed to cover each of the shaded regions below it?

a) Guess answers to this question.

b) Measure each region, approximating first to the nearest square (1 square centimeter) and then to the nearest quarter square. Notice that previous classroom practice is needed to set a standard of acceptable accuracy in order to know how carefully, if at all, the square should be partitioned.

c) Be prepared to show how you·counted. Report the accuracy of your approximation of the last area by giving numbers you are sure lie on each side of the true area.

Figure 1 no doubt suggests to you a pegboard with rubber bands stretched around certain pegs to enclose each of the polygonal regions. This device, often called a _geoboard_, is perfect for learning vocabulary and studying questions of perimeter

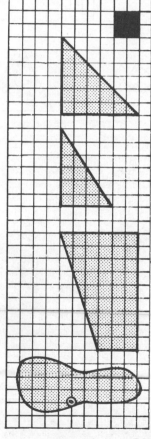

FIGURE 1

[1]Susan Roper, _Paper and Pencil Geometry_ (Pasadena, Calif.: Franklin Publications, Inc., 1966), pp. 75–78.

and area. If you have no geoboard, you may wish to set up a nail board or borrow one with at least 36 nails in it. (To make a geoboard, drive nails into the vertices of any size squares or supply yourself with holey masonite and golf tees.) You can then do the first half of Exercise 24, namely, demonstrating with a rubber band as many examples of rectangles as possible that enclose an area of six square units.

The _area of a surface_ is defined as the number of square regions needed to cover that surface. Area is a number associated with a region. For example, ten squares are needed to cover the rectangle in Figure 2a, and six square units are needed to cover the right triangle in Figure 2b. The shaded tetrahedron in Figure 2c measures 32 square inches if it takes 8 square inches to cover each of its 4 triangular regions. ⊕

As you can see by judging the area of the right triangle in Figure 2b, some surfaces have exact areas that can be found. Many, however, are, at best, approximated. The last region in Figure 1 is one such region; there is no way of cutting up squares to get a perfect fit. Archimedes was intrigued by problems having to do with the area and volume of irregularly shaped regions like this one. He devised two methods of approximating such areas, each of which suggests many interesting classroom experiences.

²Nuffield Mathematics Project, _Shape and Size 2_ (London: John Murray [Publishers] Ltd., 1968; London: W. & R. Chambers Ltd., 1968; New York: John Wiley & Sons, Inc., 1968), p. 77, by permission of the publishers.

FIGURE 2

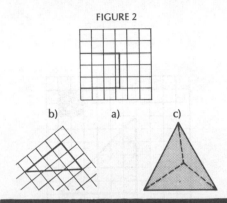

b)   a)   c)

⊕ Notes to Prospective Teachers:

The art of approximation is ready for development in youngsters and leads to adult concepts. The Nuffield project² suggests the following activities for introducing the concept of area to children in grade school. The activities are applicable to adults. Without a sense of area, even he who knows formulas is lost in new situations.

Draw around the sole of your shoe on the squared paper. When your whole group has done this, find out whose shoe covers the biggest surface on the paper.

Draw an animal on the plain side of paper. (Provide paper with squares on one side only.) Cut it out. Turn it over and compare it with the animals of the others in your group. Find out who has the animal with the biggest area. Arrange all cutouts in order of area, biggest first, smallest last. Write about what you did in your group.

(Provide a chalk box and an unopened tuna can.) See if you can find which of these has the larger outside surface area. You may use the squared paper to help you.

Gee, O Board What Secrets Wilst Thou Reveal unto Us Today?

In his book *The Method*,[3] Archimedes tells how he found such areas by analogy to physical laws. (Many of his proofs about area were suggested by cutting and placing pieces of uniformly thick material according to observed principles of mechanics.)

Archimedes cut and moved strips of material, depending on knowledge of weight and centroid. So there is likely truth in the report that he approximated the area of irregularly shaped regions in the following way: making a duplicate of the region out of a uniformly thick material and cutting out of the same material a 1-unit square, weighing them and calculating area by comparison. The ratio,

$$\frac{\text{weight of whole pattern}}{\text{weight of unit}}$$

gave him approximately the area of the irregularly shaped region in terms of the unit area. A postal scale and pieces of a cardboard box will show fair results.

Another form of approximation used by Archimedes consists of successive measurement using smaller and smaller units. His system involves successive countings of the squares within the boundary, the accuracy increasing as smaller squares are used. To get a feel for the process, look at Figure 3 in which we shade the inside squares one way, and

FIGURE 3

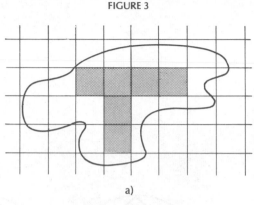

a)

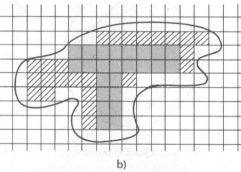

b)

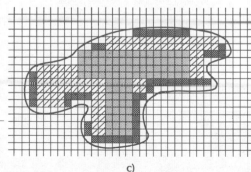

c)

the outside squares another way. Squares that lie on the boundary are white, so that the true area is less than the number of green squares plus white squares. (True area, A < green + white.) In Figure 3, the area shaded within the boundary increases as we take divisions of centimeters, half centimeters, and quarter centimeters. Counting only square centimeters in our first approximation, we find the area to be approximately 6 square centimeters.

Then in Figure 3*b*, we see a better approximation of the region as 12 square centimeters, the previous 6 plus the newly added 24 squares, $\frac{1}{2}$ centimeter on a side. (6 square centimeters). The accuracy of our approximation improves yet again when, in Figure 3*c*, we count within the boundary all squares of $\frac{1}{4}$ centimeter on a side. If we had standard measurements in mind, we might take an even better approximation by counting square millimeters, 100 of them in each square centimeter. Adding the count of the square millimeters to that of the square centimeters would make a vastly improved approximation over those obtained using square centimeters or square quarter centimeters. Of course, the approximating process just described may go on and on. By using smaller and smaller units, a better and better approximation may be determined. If you wish to know the accuracy of your approximation at any point, you can count the squares on the boundary and calculate

Number of green squares < true area A < number of green and white squares.

[3]Archimedes, *The Method*, ed. by T. L. Heath (New York: Dover Publications, Inc., 1912).

11.2 THE CIRCLE

The idea of using a succession of numbers to approximate more and more accurately the true area of a region can be utilized in finding the area of a circular region. The method pictured in Figure 3 will do this. But so will the method of inscribing regular polygons in the circle. Since an inscribed polygon touches the circle at each of its vertices, repeatedly doubling the number of sides makes larger and larger polygons within the circle. Mathematicians make sure that for this process they choose regular polygons whose areas they can readily compute; then they can write down a succession of computed areas that approach a numerical upper limit, the area of the circle. Figure 4 shows the beginning of this process of polygonal areas approaching that of a circle with a radius of 1 inch.

Stop now and sketch the appropriate inscribed polygon in each circle in Figure 4*b* and 4*c*. We say a figure is *inscribed in* a circle when all its vertices lie on the circle. Inscribe regular polygons evolved by your repeatedly doubling the number of sides (number of vertices), starting from the shaded square region in Figure 4a. Calculate the area of the square and later you will see how to calculate the area of each of the regions you drew; meanwhile, notice how diminutive the difference rapidly becomes between the area of the polygon and that of the circle.

It was by a method similar to this that Archimedes proved that the area within a circle has to be related to its radius.

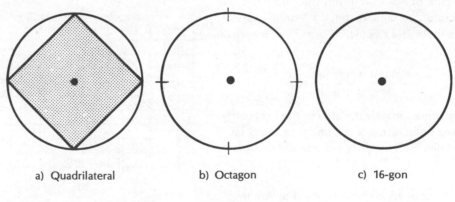

a) Quadrilateral   b) Octagon   c) 16-gon

FIGURE 4

We write $A = \pi r^2$, where r is the length of the radius.

Archimedes proved that this was the exact area by eliminating other possibilities.

Here is his reasoning, the outline of a mathematical proof that $A = \pi r^2$.

OUTLINE OF PROOF

I. Suppose the area, A, were some number, x, less than πr^2. This cannot be since Archimedes could display a polygon inscribed in the circle and bounding an area greater than x. He could do this for each number x; so $A \geq \pi r^2$.

II. Similarly, he showed it impossible for the area within the circle to be a number, X, greater than πr^2; for each number X he could circumscribe a polygon bounding area less than X. So $A \leq \pi r^2$. Thus, by elimination as shown in I and II, the area within the circle must be equal to πr^2.

It is even easier to see why πr^2 was thought to be the area if you look at a sequence of pictures in an NCTM source book.[4] There you will see the two half regions of the circle unrolled and redistributed so that the total looks very like a parallelogram of base πr and height r.

---

[4]National Council of Teachers of Mathematics, *Measurement, Topics in Mathematics for Elementary School Teachers*, Booklet 15, 1968, pp. 26–27.

♪ Note to Prospective Teacher.

It is a great opportunity, for in the seeking you will be less likely to parrot standard formulas and by so doing miss the learning experience. This text is designed for learning via involvement, even if that learning follows a period in which you feel beleaguered by questions. So let it happen! There is yet another reason for courting unusual and basic questions: this is a much better time to think than when you are on your feet in front of a class or you have one night to figure out something alone. It is becoming more important to know why; it appears that children will continue to be less and less accepting of "teacher's word" on things. For instance, a few years ago you were in the majority if you took "teacher's word" that the area enclosed by a circle measured exactly πr^2; today you would be in the minority. That the area of a circle is πr^2 is really far from obvious, isn't it? You must remember that a person's unquestioning acceptance of such a formula stems from years of indoctrination (brainwashing?) and classroom habits different from those of many present-day children. Let's hope that curiosity and unthreatened questioning are continually nurtured by our schools, and let us set about preparing ourselves for classrooms a bit different from those we knew.

EXERCISE 5:

If a man wanted to make a brick patio the shape of a semicircle against the front wall of his house,

a) what radius should he use to sketch the outline, if he knows he wants a patio of about 25 square feet? (Find radius to the nearest half foot; thus 22/7 and 3.14 are sufficiently accurate approximations of π for this problem.)

b) what area will the patio cover if he uses all of a 16-foot wall for the straight edge of his patio?

c) what length of trim must he buy for the curved edge of the patio built against a 16-foot wall? (Watch it!)

d) Does the trim cover the surface of the patio? Are you calculating area in answering Exercise 5c?

11.3 AREAS OF POLYGONS AND COMPOSITE SURFACES

Even after the preceding discussion, you may have some misconceptions or uncertainties about the concept of area. Here are several questions designed to help you become conscious of your needs. ◑ An attempt to answer each one fully and rationally will start you on your way to a correct study of this section.

FIRST QUESTION:

Can Mr. Jones give his cows more grazing grass by arranging his buildings in a cluster (Figure 5a) rather than apart (Figure 5b)?

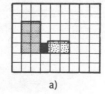

a)

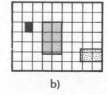

b)

FIGURE 5

SECOND QUESTION:

A common classroom might have dimensions of 20 feet by 20 feet and hold 4 rows of desks. Can you make a scale drawing of a hypothetical room of the same area, but of such a shape that no one could put in two rows of children's desks?

THIRD QUESTION:

Can an 8 by 8 square be cut and reassembled, as Figure 6 indicates, into a triangle having a base of 10 units and a height of 13?

FOURTH QUESTION:

Figure 7 shows two side views of the same pack of cards.

Is the area of the side view (the edges) smaller when the cards are stacked so that they have a rectangular side view (Figure 7a) than when the deck is slanted as in Figure 7b?

The rectangular area of the deck's side is the same as the area of a parallelogram when the deck is slanted. Each parallelogram has a related rectangle. The area within a rectangle is so easy to compute that it pays to develop for yourself a system by which you find the related rectangle for each parallelogram. Give yourself every opportunity to practice when you see a parallelogram and pretend you are watching the side of a slanted deck of cards as you straighten them. In each case, the base of the parallelogram is the same as the base of the rectangle, and there is a measurement of the

FIGURE 6

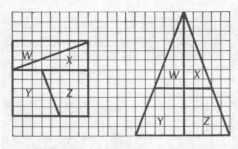

a)

FIGURE 7

a)   b)

parallelogram that is the same as the height of the rectangle.

In Figure 2a, you saw an example of how the area of a rectangle is computed from the lengths of two sides. When the long side and the short measure as rational numbers, it is obvious that area A equals the length times the width. If you had a larger vocabulary, a mathematician could prove to you that the area bounded by a rectangle is numerically the same as the measure of the long side multiplied by the measure of the short side—even in the case where one or both of the dimensions is an irrational number.

$$A = l \cdot w$$

is written shorthand for this fact.

EXERCISE 6:

Copy the side view of the deck of cards in **Figure 7b**, and then sketch in, with dotted lines, its appearance when the deck has been straightened without moving its bottom card. Then compute the area of the side of the deck if each card is 3 inches long and .02 inches thick.

EXERCISE 7:

A parallelogram can be thought of as a rectangle with a corner cut off and moved to another place. Draw a rectangle of 2 inches by 3 inches; then draw the parallelogram you see after a cutting and moving, using dotted lines to indicate the old outline of the rectangle.

Since the area of a rectangular region equals $l \cdot w$, the area of any parallelogram is the $l \cdot w$ of its related rectangle—an obvious relation when you think of the side of a deck of cards or a region containing moveable parts. Hence,

$$A_{rectangle} = l \cdot w \text{ implies } A_{parallelogram} = l \cdot h,$$

where $h = w$ in the related rectangle.

A triangle is also related to a certain rectangle via a parallelogram that the triangle would divide in half. This way of computing the area of a triangle appeals to common sense. Let us look at it. The method of counting squares breaks down for any other than right triangles (which accomodate the squares on two sides); so let us leave behind the method of counting squares inside a triangle. In Figure 8, we see three physical movements demonstrated, each of which stresses the relation of a triangular region to that of a rectangle or a general parallelogram. To understand them fully, imitate, in your imagination or with paper cutouts, each movement and its reverse in an attempt to catch as wide an understanding of their relation as possible. Some relationships may be true for any parallelogram. The following exercises should help you refine your ideas of the relationships demonstrated in Figure 8.

EXERCISE 8:

Figure 9 is a triangular region.

a) Trace Figure 9 and sketch in the dotted lines of a related parallelogram that embodies the idea of

Cut on the Diagonal

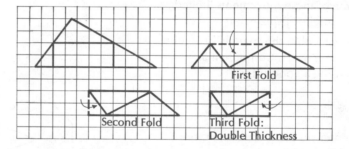

First Fold

Second Fold

Third Fold: Double Thickness

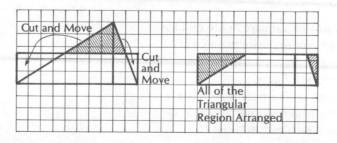

Cut and Move

Cut and Move

All of the Triangular Region Arranged

FIGURE 8

Figure 8a—if there is such a parallelogram.

b)   Similarly, retrace Figure 9 and try to sketch in a rectangle according to the principle used in Figure 8b.

c)   Can you use the principle of Figure 8c to express the area of the triangle in Figure 9? If so, demonstrate in some manner using cutouts on paper.

d)   Can you use each of the three maneuvers in Figure 8 to find the area of the triangular region of Figure 9, or are some of these movements limited to special kinds of triangles? Explain.

e)   Compute the area within the triangle of Figure 9, using one of the maneuvers. Compute it as closely as you can, using a good ruler.

EXERCISE 9:

Attempt to write sentences for each of the three visual statements of Figure 8. It may help to start each sentence "Area within a triangle equals . . . "

EXERCISE 10:

a)   Assuming each side of a square in the grid has length of $\frac{1}{2}$ centimeter, approximate the area of each triangular region in Figure 8. Stick to the method suggested by each figure.

b)   By two of the three methods demonstrated in Figure 8, find the area within the right triangle in Figure 2b. Write and sketch with sufficient accuracy so that each method can be followed.

The demonstration in Figure 8c involved cutting

FIGURE 9

An Octahedron as the
Union of Eight Triangles

and moving pieces. The mathematical concept allowing this affirms that the area enclosed by congruent polygons is the same regardless of where the polygons are placed. This is the concept which makes one sure that Mr. Jones cannot decrease nor increase the grazing area on his farm by moving his buildings around. This mental or actual cutting is an invaluable tool in calculating the area bounded by all sorts of polygons—even "horrendous" ones. With this tool you will be able to reject or affirm the correctness of shortcut formulas and devise your own.

Now you can develop an approach to finding the areas of the two regular polygons you inscribed in a 1-inch circle, (Figures 4b and 4c). Because they are regular, they can be seen as the union of adjoining congruent triangles meeting like pie slices at the center of the circle. An octahedronal region has area 8 ($\frac{1}{2}b \cdot h$), where b is the length of a segment of the octahedron and h is the triangle's height drawn to this base. Although trigonometry is needed to find the base and height of the polygons in Figure 4, you can get adequate practice splitting polygons into triangles in Exercises 20–22 and in the following exercise.

EXERCISE 11:

Complete the table of measures for the interior of a regular polygon.

No. of sides	Measure of each side	Distance from center to side	Measure of area
5	4	$2\frac{3}{4}$	
6	3	2.6	
8	5.2	7.8	

Many records, dating back thousands of years, show that, without a doubt, the Egyptians and Archimedes were able to calculate the area of many figures and the volume of a good number of 3-dimensional shapes. They developed formulas for calculating these measures, but in many cases their work was inaccurate.

One instance of an inaccurate shortcut was the Egyptian calculation for area within an isosceles triangle (one having lengths b and c equal); they made up a formula, $A = \frac{1}{2}a \cdot c$, which is not true for the triangle in Figure 10. A count of squares shows the triangle's area to be 12 squares, yet by the Egyptian calculation,

$$A = \frac{1}{2}a \cdot c$$

$$= \frac{1}{2}6 \cdot 5$$

$$= 15, \text{ instead of } 12.$$

Although that was a false formula the Egyptians used,

$$A = \frac{1}{2}ac$$

is correct for some triangle.

FIGURE 10

EXERCISE 12:

a) Does $A = \frac{1}{2}ac$ hold true for any isosceles triangle where $b = c$?

b) Draw a triangle for which $A = \frac{1}{2}ac$, where a and c are lengths of two sides.

c) Draw a generalization about all triangles where $A = \frac{1}{2}ac$.

Check out, for yourself, two Egyptian formulas that ought to be consistent, since they are both about quadrilaterals. The Egyptians thought that the area, A, of any quadrilateral could be computed as $A = \left(\frac{a + c}{2}\right)\left(\frac{b + d}{2}\right)$, where a and c are lengths of opposite sides. Another Egyptian formula calculated the area within a trapezoid (a quadrilateral with two parallel sides); they used $A = h\left(\frac{b_1 + b_2}{2}\right)$, where b_1 and b_2 represent lengths of the parallel sides and h is the distance between b_1 and b_2.

EXERCISE 13:

Refute or confirm each of the formulas. If you judge a formula to be sensible, show it to be so by cutting and rearranging the pieces of the regions.[5] Continue your investigation to cover Exercises 14 and 15.

[5] Archimedes did not guess at formulas but developed them logically, often by noting the physical parallel of cutting and rearranging as we did with the parallelogram.

EXERCISE 14:

Is the Egyptian shorthand way of calculating area as $\left(\dfrac{a+c}{2}\right)\left(\dfrac{b+d}{2}\right)$ correct for any and all quadrilaterals? Sketch all the shapes for which the formula works.

EXERCISE 15:

Is the area of a trapezoid related to its height and the length of both its parallel sides? (*Height* of a trapezoid is the distance from the base to its parallel side, measured along a line segment perpendicular to the base.)

Is $A = \dfrac{h}{2}(b_1 + b_2)$ correct?

Are there any shapes for which area is both $A = \dfrac{h}{2}(b_1 + b_2)$ and $\left(\dfrac{a+c}{2}\right)\left(\dfrac{b+d}{2}\right)$? Explain.

Finding surface area of various cylinders, cones, and pyramids amounts to a sleuth game as satisfying as many of those marketed. Let us define these closed surfaces so that we can communicate in mathematical language. A *cylinder* is a surface composed of two parallel pieces of planes and a cylindrical surface between them; a cylindrical surface is traced out around a set of parallel lines that intersect both planes. (See Figure 11.) If the curve traced by the parallel lines is a polygon, we call the closed surface a *prism* (Figure 11a); if it is any other simple closed curve, we merely call the closed surface a *cylinder*. A prism is a special cylinder.

The *pyramid* and the *cone* are also closed surfaces; however, their sides draw into a point as does an Egyptian pyramid or a tepee. (Figure 12). A *cone* is that closed surface composed of a piece of a plane and a conic surface meeting that plane; a conic surface is traced around a set of lines connecting a point to a plane not containing the point. (See Figure 12). If the curve traced by the conic surface is a polygon, we call that closed surface a *pyramid* (Figure 12a); if it is any other simple closed curve, we merely call the closed surface a *cone*. A pyramid is a special cone. Vocabulary and a more leisurely explanation are found in many other texts.[6]

Classification of some surfaces

Appearance	Name associated with polygonal base	Name associated with any other simple closed curve as base
Two parallel bases	Prism	Cylinder
Base and a point	Pyramid	Cone

FIGURE 11

a) Triangular Prism

b) Circular Cylinder

c) Right Cylinder

Watch for hints in the way that a problem is worded to be sure that you know which concept of cylinder or cone is being used, for these words also denote the infinite surfaces we know as cylindrical surfaces and conic surfaces. You will never be asked to find the area of either, since they are infinite in extent.

Now, back to the sleuthing of area of three-dimensional surfaces such as we have just defined. Notice

[6] Here is one text: Grace A. Bush and John E. Young, *Geometry for Elementary Teachers*, San Francisco: Holden-Day, Inc., 1971), pp. 111–6.

that we can use what we know of the area of such surfaces as rectangles, triangles, circles, to find the area of complicated surfaces. We shall start our sleuthing with an easy one that asks about a right circular cylinder with one base missing and a right square prism with one base missing.

EXERCISE 16:

a)   Which is greater, the outside area of a topless tin can 3 inches across and 4 inches high or a topless box having square bottom 3 inches on a side and a height of 4 inches?

b)   What is the total outside surface area of each?

EXERCISE 17:

To find the lateral area of the tin can in Exercise 16, did you mentally cut and unroll the side of the tin can as in Figure 13? It is a good way.

a)   Label the length and width of the unrolled tin can in Figure 13 assuming that it has radius, r, and height, h.

b)   Complete the sentences below the drawings.

c)   Apply your ideas in finding the total surface area of a right cylinder of height 10 centimeters whose bases each have an area of 47 square centimeters and a perimeter of 40 centimeters.

You will find other similar exercises to sleuth at the end of this chapter. A little observation and deduction will speed your finding of area, for all sides of prisms and pyramids are rectangles and triangles, respectively.

FIGURE 12

a) Pentagonal Pyramid

b) Cone

c) Circular Cone

FIGURE 13

The Lateral Area (Area of Sides)
Is _____ Times _____ .
The Total Area of This Closed Cylinder
Is _____ Plus _____ .

11.4 THE PYTHAGOREAN THEOREM

We need to study relationships between the parts of a triangle in order to progress further in finding surface areas, for many surfaces include triangular regions. Note that in Figure 2c, you could not compute the surface of the tetrahedron until you were told the area of each of the four congruent triangular regions. However, it is easy to compute the area within a triangle ($A = \frac{1}{2} bh$), providing you can find the height connected with a base of your choosing. Since h is by definition measured perpendicular to the base, we need a theorem about the relationship of the sides of a right triangle— "The Pythagorean Theorem."

THEOREM (PYTHAGORAS)

In a right triangle, $a^2 + b^2 = c^2$, where a and b are lengths of the short sides and c is the length of the side opposite the right angle.

Pythagoras, though he was brilliant, did not arrive at this generalization for all right triangles without any previously known fact to lead his thought in that direction. The people of ancient Egypt (about 3000 B.C.) found that they could always get a right triangle if three people pulled on certain knots of a closed loop of rope 12 units long. Three people served as vertices of a right triangle having sides of 3, 4, and 5 units.

Pythagoras was a Greek living around 500 B.C. who was not only a mathematician, but also the head

of a famous mystical brotherhood that later became alarmed over a result of his mathematics.
Here's the story! Pythagoras very likely arrived at
this theorem merely by observing a pattern (Figure
14) on the floor. (Count to verify that the number
of small triangles in the large square is equal to the
sum of the areas within squares A and B built on
the short sides.) Judging from the later necessity
of a less obvious proof, his first proof of his theorem was not mathematically sound; it may have
rested only on the intuitive flash of a floor pattern. In any event, Pythagoras' theorem, when
applied to a right triangle having short sides each 1
unit long, indicated that the long side of the right
triangle (the hypotenuse) measured $\sqrt{2}$ units. But
$\sqrt{2}$ was a new number. Up to then every number
had been expressible as a ratio of two integers. The
Pythagorean theorem was soon proven in a manner
the brotherhood had to accept. Then the poor Pythagoreans were stuck with logically recognizing an
outlandish number that represented to them a
flaw in the universe; the numerical aspects of their
philosophy were mostly abandoned, although some
contributions of the brotherhood are still with us
today.

EXERCISE 18:

What Pythagoras may have noted in the tile pattern
on the floor should suggest to you an experiment
with paper and scissors that can be tried on any
triangle having unmeasured sides.

What comparison do you notice between the sizes
of the squares built on the three sides of a triangle
that is "not quite a right triangle"?

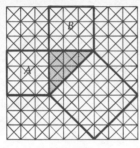

FIGURE 14

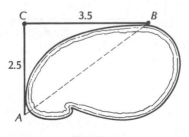

FIGURE 15

And what is true when it is a right triangle?

Now write up complete instructions on how to
determine which triangles are right triangles without use of a ruler. ◑

The Pythagorean theorem guarantees that the
length of the third side of a right triangle can always
be found. Hence we can always find the altitude of
triangles like the isosceles described in Exercise 12
and we can find the distance "as the crow flies" between two places. For the latter application, view
Figure 15 and find the distance as the crow flies between two places, A and B, if it is 2.5 miles from
A to C and 3.5 miles from C to B? We calculate,
$m(\overline{AB}) = \sqrt{(2.5)^2 + (3.5)^2} = \sqrt{18.5} \cong 4.2$ miles.

EXERCISE 19:

What is the length of the hypotenuse of the right
triangle having two sides of length

a)   1 and $\sqrt{3}$

b) $\dfrac{a}{2}$ and $\dfrac{a}{2} \cdot \sqrt{3}$

c)   4 and 4

d)   18 and 24

e)   5 and 12

Several hundred proofs of the Pythagorean theorem
have been given by a variety of people over the
ages—even President Garfield. Two proofs will
follow, the second of which doesn't even depend
upon the concept of area. The first proof rests on
the fact that the area of the center square, c^2, plus

that of the four triangles equals the area of the resulting big square in Figure 16. The proof is algebraically expressed in the space below Figure 16.

PROOF (This one relating to Figure 17)

Construct the perpendicular from C to $\overline{AB}$. Mark D on the spot where it intersects AB. By facts of similarity, $\triangle ACD \sim \triangle ABC$ and $\triangle CBD \sim \triangle ABC$.

Thus $\dfrac{m(\overline{AD})}{m(\overline{AC})} = \dfrac{m(\overline{AC})}{m(\overline{AB})}$ and $\dfrac{m(\overline{BD})}{m(\overline{BC})} = \dfrac{m(\overline{BC})}{m(\overline{AB})}$

Also, $m(\overline{AB})^2 = m(\overline{AB}) \cdot \left[m(\overline{AD}) + m(\overline{BD}) \right]$

Substituting from the first equations,

$$m(\overline{AB})^2 = m(\overline{AB}) \left[\frac{m(\overline{AC})^2}{m(\overline{AB})} + \frac{m(\overline{BC})^2}{m(\overline{AB})} \right]$$

$$= m(\overline{AC})^2 + m(\overline{BC})^2$$

Many proofs of the Pythagorean theorem are educational puzzles. Gary Hall describes in his article, "A Pythagorean Puzzle," one such puzzle using colored shapes.[7] Another variety rests on dissecting the square built on the hypotenuse by cutting on certain lines, and then reassembling the pieces so that they fill the squares built on the other sides. This amounts to a proof by congruent regions, since translation, rotation, and reflection do not change the shape and area of any region. (See Figure 18.)

FIGURE 16

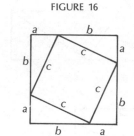

$$c^2 + 4(\tfrac{1}{2}ab) = (a + b)^2$$
$$c^2 + 2ab = a^2 + 2ab + b^2$$
$$\text{So } c^2 = a^2 + b^2$$

FIGURE 17

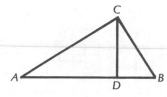

FIGURE 18

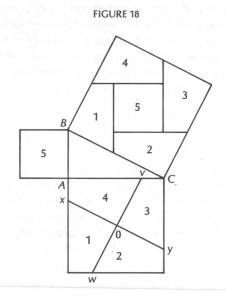

---

[7] Gary D. Hall, "A Pythagorean Puzzle," *The Arithmetic Teacher* (January, 1972), pp. 67–70.

Two other dissection proofs were devised years ago, one of them before 900. They are based on the Figures 19a and 19b, respectively, and the proofs using them are left as an exercise.

With knowledge of the Pythagorean theorem, you can investigate odd-shaped regions. A call on your creativity! When all else fails, with scissors or with a pencil, subdivide the region into regions for which you know the area. (Some ways of cutting will make the computing easier than others.) Note that the polygonal regions in Figure 20 have been sub-divided for you. Looking at each region as a union of three regions, you should be able to quickly confirm that their areas are 56 and 94 square units, respectively. For all the problems in this chapter, we can get by with knowledge of lengths in a right triangle. (Yet if it were necessary to find the length of a side of a triangle that is not a right triangle, a quick lesson in trigonometry could be given so that use could be made of the direct relationship be-tween the measure of an angle and a ratio of lengths in the triangle.)

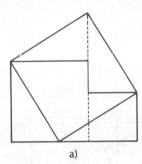

a)

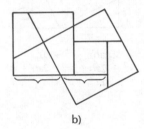

b)

FIGURE 19

11.5 EXERCISES FOR CHAPTER 11

20. The area of a regular octagon (8 sides) whose sides are all 1 inch long can be found,—and without trigonometry, if you are told that the distance between midpoints of opposite seg-ments is $1 + 2\left(\dfrac{1}{\sqrt{2}}\right)$ inch.

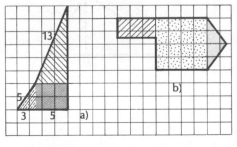

FIGURE 20

21. Find the area of a regular pentagon, each side being 1 inch long. You may want to use the fact that any two congruent triangles can be put together to form a parallelogram, or the fact that five congruent triangles can be formed from the center of a regular pentagon. To obviate the necessity of your using trigono-metry, you need to be told that the radius of the circumscribed circle is approximately .85.

22. a) Find the area of a regular hexagon in-scribed in a circle with a radius of 12 centi-meters. (Hint: Consider the hexagon as a union of triangles.)

b) Find the area of the region inside this circle and outside of this hexagon.

23. a) Find the height and area of an equilateral triangle whose side is *b* units long.

b) Find the area inside an equilateral triangle inscribed in a circle with a radius of 8 inches.

24. The area bounded by an equilateral triangle is $12\sqrt{3}$ square units. What are the dimensions of the base and height, and what is the peri-meter of this triangle?

25. a) Demonstrate with a rubber band on a geoboard as many rectangles as possible hav-ing an area of 6 square inches.

b) Demonstrate on the geoboard some tri-angles having an area of 6 square inches. About how many are there if you count only one of each set of congruent triangles?

26. In Figure 21 we have a scale drawing of a soiled section of someone's living room carpet.

    a) Make an approximation of the area of the dirty spot.

    b) Calculate the accuracy of your approximation by showing it squeezed between the inner and outer approximations of the area using square feet. (See Figure 3 to remind you of the method of inner and outer approximations.)

27. As in Exercise 26, make an approximation of the true area as wedged between inner and outer approximations of the area,

    a) using $\frac{1}{4}$ square foot

    b) using $\frac{1}{16}$ square foot.

    Shade Figure 21 so as to show one of your approximations.

28. You can find the approximate area of the soiled carpet by weighing, as Archimedes did, a copy of this carpet when cut out of some heavy homogenous substance. Try this with wood or actual carpeting.

29. a) If the diameter of a circle is doubled, what effect does that change have on the circumference? On the area enclosed?

    b) If one side of a rectangle is doubled in length, what effect does that have on the perimeter? On the area enclosed?

    c) If both dimensions of a rectangle are doubled, what effect does that have on the perimeter? On the area enclosed?

30. a) Calculate the total surface area of a salt drum, with a radius at the bottom of 2 inches and a height of 8 inches.

    b) Calculate the same for a drum, with an 8-inch radius and a height of 2 inches.

    c) Explain why a drum 8-by-2 has a different area than a drum 2-by-8.

FIGURE 21

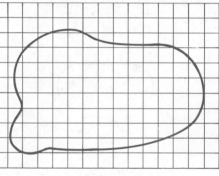

The Side of a Square Is $\frac{1}{4}$ ft in Length. One sq ft = 16 Squares.

31. Calculate the total surface area for each of the congregations of blocks in Figure 22.

32.* Make an open circular cone out of paper and level the wide end with scissors. Draw a sketch of it and its measurements. Mark or cut the paper so that you can unroll it to see the measurements of paper actually needed to make that right circular cone. Try to look at the total lateral surface in a way that helps you calculate its area; draw what you see and calculate area. (You will find that this lateral area, L, is dependent upon the slant height, s.)

33. There are numbers called square numbers numbers called triangular numbers. The angular numbers have nothing to do with but the square numbers do. Examine the squares and their respective areas in Fig to see why 1, 4, 9, 16, 25, 36, . . . and so ad infinitum, are called square numbers sixth square number is 36. State the se eighth, ninth and n^{th} square numbers

FIGURE 22

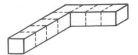

a) Each of the "Blocks"
Is a 1-ft Cube.

b) Each of the
"Blocks" Is
a Cube Whose
Edges Are 1¼ in.

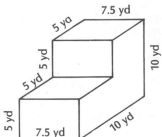

c) This Solid Is Labeled.

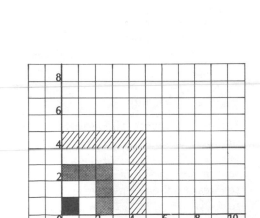

FIGURE 23

34. Find the area enclosed by each of the curves a, b, c, and d pictured in Figure 24.

35. J. S. Haldane, in his article "On Being the Right Size,"[8] reminds us of the loss of body heat corresponding to the surface area of the body. Which person's forearm loses the most heat in winter, other factors being equal—he with an arm 10 inches long and $\frac{3}{2}$ inches in radius or he with an arm 9 inches long and 2 inches in radius? (Assume the forearm is approximated by a cylinder.)

36. In the same article, Mr. Haldane reminds us that the cross-section of a bone has an area that increases with the radius of the bone.

a) Compare the areas of cross sections of a round bone with a radius of r centimeters with one having a radius of $r + 2$ centimeters.

b) The ability of a bone to support weight is dependent upon the area of its cross section. If a youngster's bone has a radius r, to what radius must it grow so that it has twice the area?

37.* There are two proofs suggested in Figure 19. Devise a proof for the Pythagorean theorem based on one of these.

FIGURE 24

a)

20 cm

b)

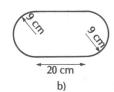

9 ft

c)

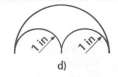

d)

---

[8] J. B. S. Haldane, "On Being the Right Size," *The World of Mathematics*, vol. 2, ed. James R. Newman (New York: Simon & Schuster, 1956), pp. 952–7.

38.* The Pythagoreans were interested in transforming an area from one rectilinear shape into another. One such problem is the finding of a triangle equal in area to that of a given polygon. There is a method by which one side at a time can be removed by construction of parallel segments and comparison of triangles formed. Can you show that a randomly chosen pentagon can be transformed into a triangle having the same area?

39. Generally, to find an area of a polyhedron, we add up the areas of the faces. Consider a right prism of seven lateral faces plus top and bottom. Must you, using a tape measure, find the areas of each of the seven lateral faces or is there a simpler method of finding lateral surface area?

40. Is there a limit to the surface area of prisms inscribed in a cylinder?

The surface area of a sphere obviously has to be less than the area of the smallest cylinder containing it. If the sphere touches at the middle of the ends and the sides of this cylinder, the area of a sphere with a radius of r is less than $2\pi r^2 + 2\pi r \cdot 2r = 6\pi r^2$. Actually, the area of a sphere is $\frac{2}{3}$ the area of this cylinder containing it, and exactly the same as the lateral area of the cylinder.

$A_{\text{sphere}} = 4\pi r^2$, where r is the radius of the sphere.

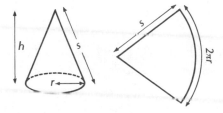

$$\frac{A_{\text{total}}}{A_{\text{sector}}} = \frac{\pi s^2}{A_{\text{sector}}} = \frac{2\pi s}{2\pi r}$$

because area varies directly with perimeter

Thus, $A_{\text{sector}} = \pi s^2 \cdot \dfrac{r}{s} = \pi rs$

Using a different approach, the lateral area of a right circular cone is found to be a sector of a circle with a radius of s, where s is the slant height. The lateral area, L, is πrs.

41. Consider the earth as a sphere with a radius of 4000 miles. If a planet had a radius of 6000 miles (1.5 times our radius), how does its surface area compare to ours?

42. Find the area of the base, the lateral area, and the total surface area for a cone with a base radius of 5 feet and a slant height of 7 feet.

ANSWERS TO EXERCISES

2. a) four lines of symmetry.

b) Only two lines of symmetry.

Note how the diagonal fails as an additional line of symmetry.

c) three lines of symmetry.

3. In this drawing, the reverse side of the paper is shaded. Fold to find the greatest diagonal. It works because your fold places the bottom edge up so that the horizontal dimension matches that of a vertical edge.

4. a) 4, 3, 8, 4.

   b) First region: 4 sq. cm., then $4\frac{1}{2}$, or $4\frac{2}{4}$ sq. cm.
      Last region: 5 sq. cm., then 25 sq. $\frac{1}{4}$ cm. or $6\frac{1}{4}$ sq. cm.

   c) No. inside < 25 sq. $\frac{1}{4}$ cm. < no. inside and on border.

      18 < 25   < 40

5. a) Area $= \dfrac{\pi r^2}{2} \cong 25$

      $r \cong 4$

   b) $32\pi \cong 100.5$ sq. ft.

   c) $C = \dfrac{\pi d}{2} \approx 25.1$ ft.

d) Trim computation is one of length.

Question 1. No, the buildings consume the same space—and take up the same area of the land—wherever they are placed.

Question 2.

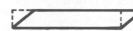

Question 3. The areas are not the same. The cuts make them look the same, but they are not. Explain why if you can.

Question 4. No, the volume of the deck of cards is invariant in a situation of no physical change.

The cards are all the same width, say 2″. So the volume of each pack is 2 times the area of the cross-section we are examining.

$V_1 = V_2 \longrightarrow$ Area$_1$ = Area$_2$

Or, another way, the sum of all the edges in pack$_1$ equals that in pack$_2$ because the length of cards and the thickness of each card, hence the pack, is the same regardless of the position they are stacked in.

6.

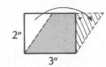   Area = 3.12 square inches

7.

8. a) On the order of Figure 8a.

   b) On the order of Figure 8b.

c) On the order of Figure 8c.

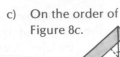

d) O.K. for all triangular regions.

e) 4/10 square inches by method of 8a.

9. b) Area within a triangle = double the area of a rectangle inscribed with height half that of the triangle.

c) Area within a triangle = the area of a rectangle with the same base and half the height.

10. a) The last picture is 25/4 square centimeters (5 centimeters by $1\frac{1}{4}$ centimeters).

b) By method a, I see half of a rectangle of 3 times 4 square units, thus 6 square units.
By method c, I see the triangle having area equal to that of a rectangle drawn at half the height along side Y and using X for base. So area = 2 times 3 = 6 square units. This checks with our definition of area as number of squares; I see 3 squares and parts of other squares—looking like approximately 6 square units.

11. 55/2, 23.4, 162.24.

12. a) The formula fails for all isosceles triangles where b and c are of equal length.

b)

c) It *will* hold for all right triangles of altitude and base represented by the lengths a and c, including the right isosceles triangle where a = c.

14. No It works for some, It is not correct for including

15. Yes; Yes; Both are true for the area of a rectangle.

16. a) box.

b) box: 57 square inches   can: 57/4 π square inches.

17. b) $LA = h(2\pi r)$
$TA = 2\pi r^2 + 2\pi rh$

c) 94 + 400 square centimeters.

18. In a right triangle, and only in a right triangle, the region built on the long side can be cut so as to fill both the square built on one short side and the square built on the other short side.

19. a) 2   b) 2   c) $4\sqrt{2}$ d) 30   e) 13

20. $2 + 2\sqrt{2}$ square inches.

21. $5\left(\frac{1}{2}(1)(.7)\right) = 1.75$ sq. in.

22. a) $A = 6(\frac{1}{2}\cdot 12 \cdot 6\sqrt{3}) = 216\sqrt{3}$ sq. in.

b) $144\pi - 216\sqrt{3}$.

23. a) $\frac{b}{2}\sqrt{3}$ is the height.

$\frac{1}{2}b \cdot \frac{b}{2}\sqrt{3}$ is the area.   b) $48\sqrt{3}$ square inches.

24. Referring to 23a we see that $12\sqrt{3} = \frac{1}{2}\cdot b \cdot \frac{b}{2}\sqrt{3}$.

So $h = 4\sqrt{3}$; h = 6; perimeter = $12\sqrt{3}$.

25. a) There are only two rectangles if you exclude the repeats (the congruent triangles).

b) On a geoboard, you can show more triangles than right triangles, and more right triangles than rectangles.

26. a) The first estimate is 5 square feet.

b) This lies between the gross measurements of square

feet inside and those on the boundary and inside combined, written 2 < 5 < 10.

27.

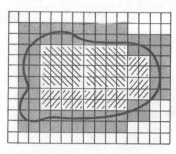

Using $\frac{1}{4}$ sq ft, My Estimate of 5 Lies Between $3\frac{1}{2}$ and 8.

The Side of a Square Is $\frac{1}{4}$ ft in Length. One sq ft = 16 Squares.

29. a) The circumference doubles and the area quadruples.

b) $P = 2L + 2w = 2(2l) + 2w \neq k(2l + 2w)$; The perimeter is greater but not twice as great.

   $A = L \cdot w = 2l \cdot w = 2 \cdot lw$; The area doubles.

c) Perimeter is doubled, and area is quadrupled.

30. a) $2(\pi \cdot 4) + 8 \cdot 4\pi = 40\pi$

b) $2(\pi \cdot 64) + 2 \cdot 16\pi = 160\pi$ four times as great a surface.

c) The greater area accompanies the greatest radius because radius is squared.

31. a) 34 square feet.   b) 53 1/8 square inches.
    c) 450 square yards.

34. a) 33 sq feet, approximately.   b) $360 + 81\pi$ square centimeters

    c) $24 + 2\pi$ square feet.   d) π square inches.

35. The second arm loses the most heat, because $4\pi \cdot 9 > 3\pi \cdot 10$.

36. a) The second bone is larger by $4\pi r + 4\pi$.

b) $2\pi r^2 = \pi(\sqrt{2r})^2$; The bone must grow to approximately 1.4 times its present radius.

38.* Try constructing parallel lines in such a way that base and height of two different triangles are the same; then one can be substituted for another. It is not a trivial problem.

39. You know you are right if you have to find only two dimensions. To calculate lateral area of a prism, you need record only height and perimeter of base.

40. Area of base approaches πr^2 as number of sides is continuously increased. At the same time lateral area approaches $2\pi r \cdot h$.
The limit to the area of inscribed surfaces is the surface of the cylinder, $2\pi r^2 + 2\pi rh$.

41. It has $\frac{9}{4}$ the surface area of earth.

42. $A_{base} = \pi \cdot 5^2 = 25\pi$.
   $L = \pi \cdot 5 \cdot 7 = 35\pi$.
   $S = 35\pi + 25\pi = 60\pi$.

Abbott, Janet S., et al. *Progress Book A*. Pasadena, Calif.: Franklin Publication, Inc., 1968, pp. 57–58, 65–67.

Archimedes. *The Method*. Edited by T. L. Heath. New York: Dover Publications, Inc., 1912.

Buchman, Aaron L. "An Experimental Approach to the Pythagorean Theorem." *The Arithmetic Teacher* (February 1970), pp. 129–32.

Bush, Grace A., and Young, John E. *Geometry for Elementary Teachers*. San Francisco: Holden-Day, Inc., 1971, pp. 150, 159–72, 198–202.

Cockcroft, W. H. *Your Child and Mathematics*. New York: John Wiley & Sons, Inc., 1968.

Colter, Mary T. "Adapting the Area of a Circle to the Area of a Rectangle." *The Arithmetic Teacher*. (May 1972), pp. 404–6.

Conway, Donald, and Dreyfuss, Martin J. *Arithmetic Skills and Problem Solving*. New York: Harcourt Brace and Jovanovich, Inc., 1968, pp. 391–8.

Copeland, Richard W. *Mathematics and the Elementary Teacher*. Philadelphia: W. B. Saunders Co., 1966, pp. 267–9.

Eves, Howard W. *Fundamentals of Geometry*. Boston: Allyn & Bacon, Inc., 1969.

Haldane, J. B. S. "On Being the Right Size." *The World of Mathematics*. Vol. 2. Edited by James R. Newman. New York: Simon & Schuster, Inc., 1956 pp. 952–7.

Hall, Gary A. "A Pythagorean Puzzle." *The Arithmetic Teacher* (January 1972), pp. 67–70.

Kasner, Edward, and Newman, James R. *Mathematics and the Imagination*. New York: Simon & Schuster, Inc., 1963, pp. 343–55.

Liedtke, W., and Kieren, T. E. "Geoboard Geometry for Preschool Children." *The Arithmetic Teacher* (February 1970), pp. 123–6.

National Council of Teachers of Mathematics. *Measurement, Topics in Mathematics for Elementary School Teachers*. Booklet 15, 1968.

National Council of Teachers of Mathematics. *Informal Geometry, Topics in Mathematics for Elementary School Teachers*. Booklet 14, 1968.

National Council of Teachers of Mathematics. *Geometry, Experiences in Mathematical Discovery*. Booklet 4, 1966.

Nuffield Mathematics Project. *Shape and Size 2*. New York: John Wiley & Sons, Inc., 1968, pp. 61–67.

Ptak, Diane M. *Geometric Excursion*. Oakland County Mathematics Project. Oakland, Mich., 1970, pp. 18–20.

Roper, Susan. *Paper and Pencil Geometry*. Pasadena, Calif.: Franklin Publications, Inc., 1966, pp. 75–78.

Schaaf, William L. *Basic Concepts of Elementary Mathematics*. New York: John Wiley & Sons, Inc., 1966, pp. 315–20.

Shanks, Daniel. *Solved and Unsolved Problems in Number Theory*. Vol. 1. New York: Spartan Books, Inc., 1962, pp. 67–71, 121–27.

School Mathematics Study Group. "Intuitive Geometry." *Studies in Mathematics*. Vol. 7. Palo Alto, Calif.: Leland Stanford Jr. University, 1961, pp. 29–30.

Stubblefield, Beauregard. *An Intuitive Approach to Elementary Geometry*. Monterey, Calif.: Brooks/Cole Publishing Co., 1969, pp. 154–68, 178–88, 191–6.

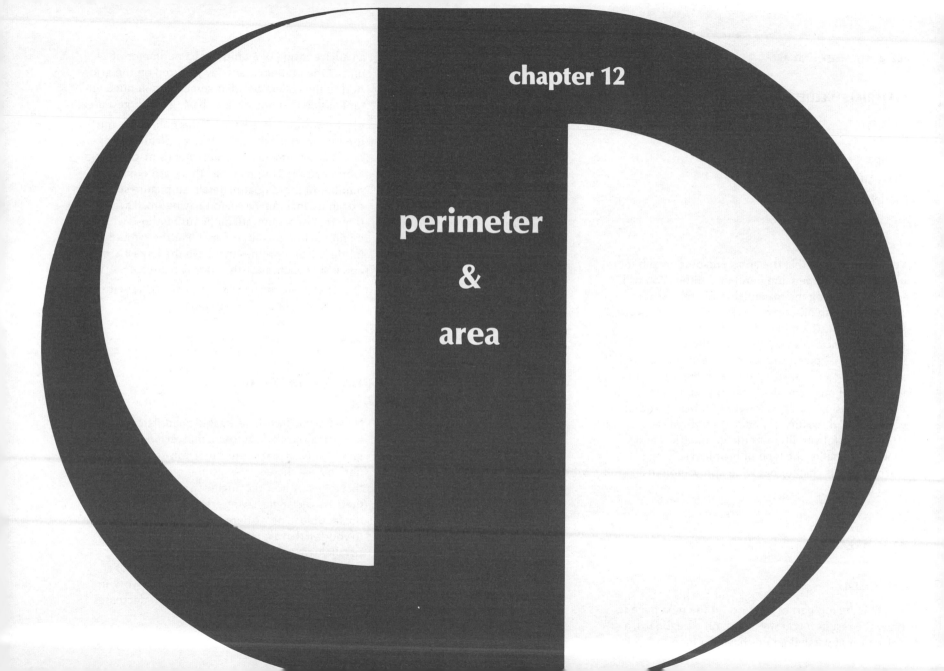

chapter 12

perimeter

&

area

MATERIALS NEEDED:

Rulers

Compasses

Plywood or heavy cardboard

Model of walls and chain

You have noticed that the areas enclosed by differing polygons of the same perimeter differ. You will find that studying the maximizing of area within a specified length of simple closed curve is well worth some thought. The minimizing of the perimeter needed to enclose a certain area is the same problem. It also has practical applications in such areas as packaging and construction. Both the know-how for practical problems, like that of finding largest enclosure using a limited length of fencing, and the geometrical discovery of pattern—both make this study pay off. In addition to the geometrical and practical benefits, the type of problem solving learned in this chapter has wide application in "real life."

12.1 AREA

EXERCISE 1:

a)   Use the top part of the grid on the next page to draw the results if you stretched, on a pegboard, a collection of parallelograms having a base of 7 units,

an adjacent side of 5 units, and a perimeter of 24 units. One of them is already pictured on the grid. Add to the collection what you can, then put a letter L inside that one or ones having the largest area.

b)   Now draw a line beneath this collection and use the bottom half of the grid for picturing pegboard arrangements of quadrilaterals made from a loop of string 24 units long. There are quite a number of these quadrilaterals, so picture some extremes, including a region having small area, then really concentrate on picturing those which might enclose the largest area. Put the symbol L inside the one or ones that have the largest area, and inside each, state the approximate area.

c)   Circle, with a colored pencil, the quadrilateral(s) of greatest area on the whole page.

12.2 THE RECTANGLE

Now that you have investigated quadrilaterals in this way, it may be obvious to you that rectangles have the greater area. It is a theorem that non-rectangles are nonoptimal shapes amongst the quadrilaterals, i.e., that for every non-rectangle of a given perimeter, there is a rectangle having greater area. That fact is not feasible to prove here, but you can readily prove to yourself that a rectangle has greater area than a parallelogram of the same base and perimeter. Watching a collapsing pipe cleaner model of a certain rectangle and remembering our work in Chapter 11, we can see that the height decreases

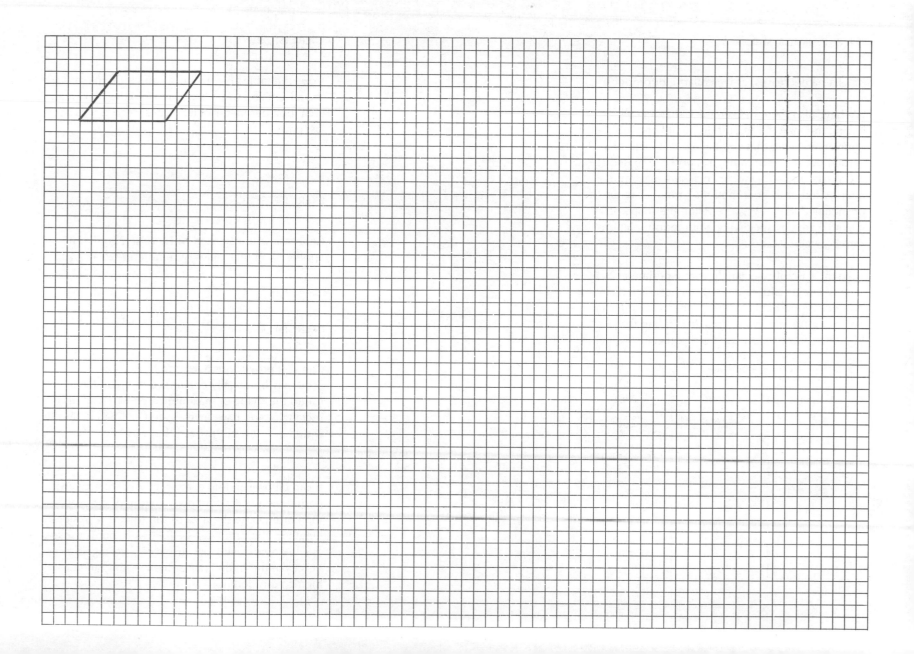

(thus area decreases) as we collapse our model.

We have also seen, in Exercise 1, that rectangles having the same perimeter do not all enclose the same area. Given that you have a rectangle of a certain perimeter in mind, it is possible to deduce the shape of the rectangle enclosing maximum area. Whether the perimeter is a number or an unspecified $2L + 2W$, you define the shape of maximal area by specifying a numerical relationship between W and L.

Take the case of a rectangle with a perimeter of 60 inches. Length plus width is 30 inches. Search among the rectangles that have this property for the shape enclosing the maximum area. You'll find that adding to the table below will help you get a "hunch."

Length	Width	Area
10	20	200

EXERCISE 2:

Copy the table above, and when you think you see the length and width that will yield the maximum area for a perimeter of 60 inches, record it as your last entry.

EXERCISE 3:

Rephrase your assertion of Exercise 2 so that it applies to all rectangles of perimeter $2L + 2W$.

EXERCISE 4:

Can you prove your assertion of Exercise 2? (You may be visually convinced that you have found the shape of maximum area when all other rectangles you have investigated can be cut and fit within, but it is not proved that way.)

It can be proven that every rectangle of a given perimeter, say 4c, confines a region of area less than or equal to the area of the square with sides of length c. The proof can be followed, but is useless as a learning device because it reflects algebraist's "tricks of the trade" rather than resting on geometric considerations alone.

THEOREM

For rectangles of a given perimeter, area is maximized in the shape of the square.

PROOF

Let $L + W = 2c$, where $2c$ is a definite number for half of the perimeter of this randomly chosen rectangle of dimensions L and W.

Then $L - c = c - W$

Let $L' = L - c$ and let $W' = c - W$

So $L' = W'$

$$A_{\text{rectangle}} = L \cdot W = (L' + c)(c - W')$$
$$= (c + L')(c - L')$$
$$= c^2 - (L')^2$$

Since $(L')^2 \geq 0$, $c^2 - (L')^2 \leq c^2$

So $A_{\text{rectangle}} \leq c^2$ or $\left(\dfrac{L + W}{2}\right)^2$

EXERCISE 5:

What is the area of the largest rectangular pasture Orville can build with 1520 yards of fencing?

What are the exact dimensions?

By the way, a little examination will convince you that holding one dimension constant and increasing another will increase both area and perimeter. Since area and perimeter are both increasing functions of length and width, the statement of the theorem just proven can be reworded magnificently as follows:

EQUIVALENT THEOREM

For rectangles of a given area, perimeter is minimized in the square.

EXERCISE 6:

Verify the equivalent theorem for yourself with various rectangles of 64 square units.

EXERCISE 7:

You are building a storeroom of area A. You want to minimize the cost of the walls.

What shape should you build it if all the walls are to be of the same material? For interesting variations on this problem, see Problem 2 at the end of the chapter.

It is important to generalize the relationship of the last theorem and its equivalent. In algebraic symbols, these two theorems are saying,

THEOREM

If $a + b$ is a constant,
then $\qquad\qquad a \cdot b$ is maximal when $a = b$.

EQUIVALENT THEOREM

If $a \cdot b$ is a constant,
then $\qquad\qquad a + b$ is minimal when $a = b$.

Seeing the theorems written this way may help you to realize applications completely devoid of the mention of squares.

EXTENSIONS OF THESE THEOREMS

If $x + 2y = c$, a constant
then $\qquad x \cdot 2y$ is maximal when $x = 2y$.

If $5x + 18y = c$,
then $\qquad 5x \cdot 18y$ is maximal when $5x = 18y$.

If $\dfrac{x + y}{2} = c$,

then $\qquad \dfrac{x}{2} \cdot \dfrac{y}{2}$ is a maximal when $\dfrac{x}{2} = \dfrac{y}{2}$.

If $x + 2y = c$,

then $\qquad xy = \dfrac{1}{2}(x)(2y)$ is maximal when $x = 2y$.

Now the equivalent theorem,

If $x \cdot 2y = c$,
then $\qquad x + 2y$ is minimal when $x = 2y$.

If $\dfrac{x}{2} \cdot \dfrac{y}{2} = c$,

then $\qquad \dfrac{x + y}{2}$ is minimal when $\dfrac{x}{2} = \dfrac{y}{2}$.

If $xy = c$,
then $\qquad x + 2y$ is minimal when $x = 2y$.

If $xy = c$,
then $\qquad 5x + 18y$ is minimal when $5x = 18y$.

EXERCISE 8:

What are the dimensions and area of the largest rectangular dog pen you can build with 60 feet of chain link fence costing $8 per yard if you are restricted to considering only a dog pen,

a) of over 10 feet in width?

b) with a width of 10 feet or less?

c) adjoining another pen on one side?

d) Which of the three is the cheapest per square foot? See Problem 1 at the end of the chapter.

12.3 THE TRIANGLE

Now let's devote some thought to triangles enclosing maximum area. This means that $\frac{1}{2}bh$ is maximal, and thus bh is at a maximum. It simplifies the problem to consider, first, those triangles for which we specify perimeter and length of one side, and then to graduate to triangles where only perimeter is given.

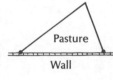

Pasture

Wall

FIGURE 1

EXERCISE 9:

a) Sketch four members of the set of triangles formed on an 8-foot side, and having a perimeter of 22 feet.

b) Circle the triangle that contains the greatest area of the four. If there is one having maximum area, make sure you include it and label it *maximal.*

EXERCISE 10:

A whole set of triangular pastures can be made from two gates hinged to certain places in a wall, depending on the lengths of the gates. (See Figure 1 for one such pair of gates.)

a)   Sketch 4 members of the set of triangles suggested by a wall and 2 gates hinged 6 meters apart on the wall, enclosing a pasture whose perimeter is 16 meters. Circle the triangle that encloses maximum area, if you think you've found it.

b)   The sum of the lengths of the 2 gates was 10 meters in Exercise 10a. Note that the answer for 10a shows those points where the third vertex could lie. Show that there is nothing magical about 10 units, the sum of the gate lengths used in Exercise 10a. You can convince yourself and others of this by drawing a wall and tracing the possible third vertices that could exist if the lengths of the gates totaled $b + c$.

Hint: Pin down a piece of string at the hinge places in a scale drawing, and trace, with a pencil, all the places where the third vertex could lie— where the gates could meet if their total length was that of the string. Mark, with a red pen, the point where the vertex lies when the triangle encloses the maximal area; for this optimum triangle, what relation has each of the gates to the length, $b + c$?

The two preceding exercises concentrate essentially on the problem of finding the triangular shape that yields maximum area for a certain perimeter and base. We are now in a position to tackle the second, and more general, problem: Of triangles of a given perimeter, which shape(s) enclose maximum area? (The base is unrestricted in this problem.) This can be discovered on the geoboard through disciplined efforts with a string tied in a simple closed curve. (Not a rubber band because

we must not change the perimeter.) There are many triangles with the same perimeter—they are *isoperimetric.* Your present task is to find those in which the dimensions maximize base times height so that $\frac{1}{2}b \cdot h$ is maximal; thinking about maximizing $b \cdot h$ should speed your experimentation with string.

EXERCISE 11:

Set up a table like that used in Exercise 2 including columns for lengths of the three sides, one of the heights, and the computed area. Fill it in in an effort to find the shape of triangle containing greatest area, remembering that you already found out that the triangle of maximal area is isosceles. The table can be filled with general dimensions or with specific numbers so long as some entries or summary sentences generalize what you find.

Our reasoning as to this shape used the fact that the maximum triangular area, with one side and perimeter specified, is that enclosed within an isosceles triangle. This fact is the stepping-stone in the proof that the triangle of maximum area for a given perimeter is the equilateral triangle.

THEOREM

Every triangle of perimeter 3p encloses a region
with an area less than or equal to the area of the
equilateral triangle having each side of length p.

PROOF

Suppose that there exists a triangle with at least
two sides of unequal lengths, b and c. (See Figure
2).

Construct an isosceles triangle on side d, making
equal sides of length (b + c)/2, to optimize the
area within the triangle built on side d.

This construction could have been done on any
side.

Hence, the only triangle whose area cannot be
further optimized is the equilateral triangle.

Because area and perimeter are both increasing
functions, the above theorem can also be read in
its equivalent form:

EQUIVALENT THEOREM

For all triangles having a given area, the equilateral
triangle has the least perimeter.

EXERCISE 12:

Roberta wishes to plant trees in a fenced triangular
grove.

a)   What is the least fencing needed for enclosing
a grove of $2000\sqrt{3}$ square meters? What shape

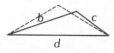

FIGURE 2

triangle should she use to minimize the fencing
needed?

b)   How much fencing is needed if she builds her
grove up against a wall $200\sqrt{3}$ meters in length?

12.4   THE ULTIMATE

Let us now turn our attention to any polygon, and
try forming some general ideas on isoperimetric
polygons of maximal area. You may wish to adopt
the mathematics vocabulary of 5-gon, 10-gon, 14-
gon, n-gon, n+1-gon, 2n-gon, etc. to ease communication. We say that something is true for any
n-gon, when we wish to indicate that it is true for
any polygon. (A reader understands the convention
that the use of n is limited to positive integers
greater than 2.) Before beginning an exposition,
try drawing a few polygons and observe a few
things in exercises below. Then we shall be on
more equal levels of communication.

EXERCISE 13:

a)   Construct a square on a piece of paper. Draw a
circle passing through all four vertices of the
square. (Use a compass in this circumscribing of a
circle.)

b)   Sketch an equilateral triangle and its circumscribed circle on graph paper. Sketch inside that
circle, using dotted lines, another triangle having
two vertices the same as the first triangle and a
third vertex somewhere else on the circle. Com-

pare their approximate areas.

c)   Repeat the steps of part *b* for a regular hexagon and a nonregular hexagon that has five of the same vertices and one other that is on the circle.

d)   State what you notice about the regular polygon.

EXERCISE 14:

Inscribe in a circle any *n*-gon. (The *n*-gon is inscribed when the vertices are all on the circle.) Now add another vertex on the circle and connect the $n+1$ vertices to form an $n+1$-gon. Which has the greatest area, and why?

EXERCISE 15:

Is there a regular polygon around which we cannot circumscribe a circle? Explain if you can.

EXERCISE 16:

Draw an equilateral triangle and its circumscribed circle on graph paper. (You may trace this from your answer to Exercise 13*b*.) Beside it, draw two other circles congruent to this circle, and inscribe regular polygons of 6 and 12 sides, respectively. The three circles then circumscribe the equilateral triangle, the regular hexagon of Exercise 13*c*, and the regular 12-sided polygon whose vertices can be located by twice tracing the hexagon's vertices. How does doubling the number of sides of a regular inscribed polygon affect the area enclosed?

You no doubt notice that maximizing area in an *n*-gon has some connection with the circle. The circle is something special. Drop a bit of oil on a slick surface—water or glass, for instance—and watch the shape it spreads to. Do we really know why? Maybe not, but physicists say that the droplet forms a circular region in keeping its inner tension by having the least perimeter in contact with the foreign material. So the droplet of area *A* spreads itself into circular form when it minimizes its perimeter. Another good way to see the maximality of the circle's area demonstrated is to blow a soap bubble between two parallel glass plates that have previously been wetted by the soap solution. As soon as an edge of a soap bubble touches a glass plate, it assumes a circular outline.[1] The fact that the bubble finally becomes a cylinder—that both edges of the soap bubble become circles—is related to the fact that the circle has the greatest area for a given perimeter. This fact is also suggested by a sleeping cat who arranges his outline as circularly as possible in order to minimize the surface over which he loses body heat; The cat has no intention of decreasing his body volume, he merely seeks a position that minimizes surface.

Viewing this from the other direction, we have seen, in the exercises, that the inscribed polygons of increasing area approached closer and closer to the circular outline. To see this most clearly, take a string knotted into a simple closed curve. Place it over a sheet of paper, observing the area within as

---

[1]Richard Courant, and Herbert Robbins, *What Is Mathematics?* (New York: Oxford University Press, 1960), p. 394

you make different polygons by holding the string taut with your fingers or with a pegboard. Tracing around the regions and comparing their areas will convince you that the more sides the polygon has and the closer its outline is to a circle, the larger its contained area. For a given perimeter, the circle is the shape containing the greatest area.

There is an obvious argument we can present to show that an n-gon of given sides has maximal area when its vertices lie on a circle. This argument originated with Jacob Steiner and can be found in Polya's excellent chapter called "The Isoperimetric Problem."[2] Here is how it goes. Inscribe any n-gon in a circle. Imagine the sections of the circular region bounding the polygon (the ones shaded in Figure 3a) to be made of plywood. Now deform this polygon by changing the angles of the polygon (see Figure 3b). After deformation, note the outlining lumpy curve that is a collection of circular arcs having a total perimeter equal to that of the circle.

We know that among curves of equal perimeter, the circle contains the maximum area, and we know that the plywood sections have not changed their area, so we can subtract the plywood sections and calculate that the n-gon in Figure 3a is the n-gon of greatest area.

FIGURE 3

a) An Inscribed Polygon

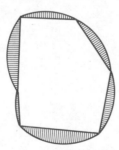

b) Flexible Joints and Plywood Segments

---

[2]George Polya, *Induction and Analogy in Mathematics, Mathematics and Plausible Reasoning*, vol. 1 (Princeton, N.J.: Princeton University Press, 1954), pp. 168–89.

REASON

Area of n-gon in Figure 3a + area of sectors > area of n-gon in Figure 3b + area of same sectors.

We can also prove that a regular n-gon has the greatest area of all n-gons of a given perimeter. Consider any n-gon inscribed in a circle. (See Figure 4a.) Consider any two adjacent sides of this n-gon and the chord across the n-gon that completes a triangle. (See Figure 4b.)

We know that for the perimeter to stay fixed, the sum of the lengths of these two sides must stay fixed. But the maximal area in a triangle built on a fixed chord is attained when the triangle is isosceles; so the two adjacent sides of the n-gon should have equal length. A similar argument applies to all pairs of adjacent sides. Hence, the regular n-gon of a given perimeter has the maximum area.

Applications of the special properties of the circle and regular polygons are many and intriguing. Interesting applications are in the problem section at the end of the chapter and a few exercises are below.

EXERCISE 17:

a)   Approximate the greatest area you can enclose with chicken wire 31 feet long? What outline does the fence follow along the ground?

b)   What is the area of the largest region you can enclose with chicken wire 31 feet long stretched taut between 4 posts? How about between 8 posts?

1. a) What is the maximum rectangular area that can be fenced with P units of fencing material?

   b) What are the dimensions of the maximum rectangular pasture using P units of fencing and built against a wall? Remember that half of a square is bigger than half of any other rectangle, and that you can make a square by reflection of a certain rectangle across the wall.

2. You are building a storeroom of area A, and you want to minimize the cost.

   a) What dimensions should you plan if one pair of opposite walls uses boards costing twice as much as the other?

   b) What dimensions if one pair costs three times as much as the other?

   c) What dimensions if one pair costs 15 times as much as the other?

3. Discuss the question of whether or not there is any minimum area that can be enclosed by a rectangular fence with a length of P units.

4. Discuss the question of whether or not there is any maximum perimeter needed to enclose a yard of 60 square meters.

5. a) If a fence has its ends on a straight wall, what is the most efficient shape for enclosing a pasture?

   b) If you have five straight pieces of fencing, how can you place the pieces so you can make an enclosure containing the most area? (Describe the relative positions of the five vertices.)

6. In Figure 5, you see two nonparallel walls connected by a loose piece of chain fence having a length of L, long enough to reach between the two walls. The Park Service wishes to consider different options for fencing off an exhibition from the touch of tourists. The chain fence can be placed in any configuration by means of metal rods plunged into the earth. They wish to enclose a maximum area.

   a) In what shape should they arrange a chain fence going between the two walls if they fasten two ends to points an equal distance out from the corner?

   b) In what shape should they arrange the chain if the ends are attached to randomly chosen points on each wall?

   c) If the chain is fastened on the upper wall, where should they hook the other end of the chain and in what shape should they mold it so as to provide an enclosure of maximum area? (To keep your picture intuitively obvious, use a chain with a length somewhere between the shortest that will reach and that shown in Figure 5.)

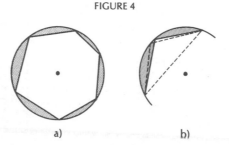

FIGURE 4

a)   b)

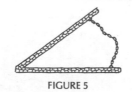

FIGURE 5

ANSWERS TO EXERCISES

1. a) Here is one:

h) Here is one:

2. 15 by 15 yields maximum area.

3. $L = W$, so the square is the rectangle shape of maximum area.

4. Think of something algebraic derived from maximizing the product of x and $30 - x$.

5. $l + w = 760$ yards.
    If $l = w$, then $l = 380$ yards
            and $w = 380$ yards,

    $A_{max.} = (380)^2$

6.

Perimeter $= 2 \cdot 4 + 2 \cdot 16 = 40$ $P = 2 \cdot 8 + 2 \cdot 8 = 32$

7. Minimal cost if $l = w$; build a square room.

8. a) biggest is a square pen of 15 feet on a side; $A = 225$ square feet.
    b) width of 10 feet and length of 20 feet; $A = 200$ square feet.

c) Note two shapes where $l + 2w = 60$, amongst many such.

Since square is maximal for rectangles, half a square should have more area than half of any other rectangle.

Half a square, so $l = 2w$
    $l + 2w = 60$
Thus $4w = 60$
Hence $w = 15$ and $l = 30$.

d) Total cost of 20 times $8. = $160. Fencing best used by method c: 450 square feet for $160. = 35¢ square foot.

9. a) Here are some

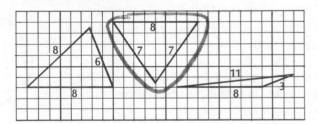

10. a)

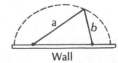

Wall

gate a + gate $b = 10$ meters
            $a + b = 10$
Here is a tracing of *all* places where the third vertex could lie when $a + b = 10$.

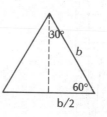

Maximum Area

Wall

b) The triangle of maximum area is the one having the greatest height, the one at the top of the semiellipse.

$$2000\sqrt{3} = \tfrac{1}{2}bh = \tfrac{1}{2}\left(\frac{b\sqrt{3}}{2}\right),$$

since maximum area is inside an equilateral triangle.

$$2000\sqrt{3} = \frac{b^2\sqrt{3}}{4}$$

$$40\sqrt{5} = b$$

11

Base	Side		Height	Area of a triangle	Observations
d	b	c			
6	b	b	$\sqrt{b^2 - 9}$	$\tfrac{1}{2}6 \cdot \sqrt{b^2 - 9}$	
10	b	b	$\sqrt{b^2 - 25}$	$\tfrac{1}{2} \cdot 10 \cdot \sqrt{b^2 - 25}$	
6	10	10	$\sqrt{81}$	$\tfrac{1}{2} \cdot 6 \cdot 9 = 27$	P = 26
4	7	7	$\sqrt{45} = 3\sqrt{5}$	$\tfrac{1}{2} \cdot 4 \cdot 3\sqrt{5} = 6\sqrt{5}$	P = 18
10	4	4	3	$\tfrac{1}{2} \cdot 10 \cdot 3 = 15$	P = 18
6	6	6	$3\sqrt{3}$	$\tfrac{1}{2} \cdot 6 \cdot 3\sqrt{3} = 9\sqrt{3}$	P = 18, max. area

12. a) minimal fencing for the area

$$h = \sqrt{b^2 - \frac{b^2}{4}} = \sqrt{\frac{3b^2}{4}} = \frac{b\sqrt{3}}{2}$$

b) $2000\sqrt{3} = \tfrac{1}{2}(200\sqrt{3})(h)$

$10 = \tfrac{1}{2}h$

$20 = h$

If h = 20,

$b^2 = 20^2 + (100\sqrt{3})^2$

$= 30400$

$b = 40\sqrt{19}$

Perimeter $= 2b = 80\sqrt{19}$ meters.

Wall

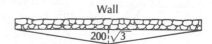

$200\sqrt{3}$

13. b)

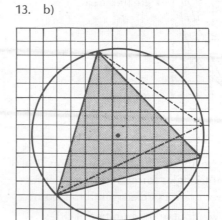

$A_{\text{equilateral triangle}}$
$\cong \tfrac{1}{2}(10.8)(9.2)$
$\cong 49.7$

$A_{\text{other triangle}}$
$\cong \tfrac{1}{2}(10.8)(8.7)$
$\cong 46.9$

$A_{ot} < A_{\text{equil.}}$

c)

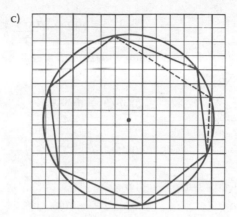

14.

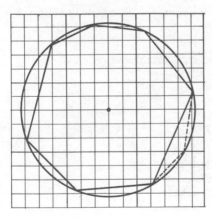

The dotted lines indicate the two new sides of the $n + 1$-gon.

The $n + 1$-gon has the greater area.

15. No, we can circumscribe about any regular polygon. The presence of congruent triangles formed from the center implies equal radii if we draw a circle around this point.

16. The area increases as we double the number of sides.

17. a) Chicken wire should follow the outline of a circle.

$2\pi r = 31$ feet

$r \cong 5$ feet

So $A \cong \pi r^2 \cong 25\pi$ square feet.

b) 4-sided regular polygon

$\frac{31}{4}$ on a side

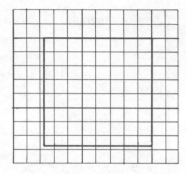

$A = \left(\frac{31}{4}\right)^2 \cong 60$ square feet

8-sided regular polygon

$\frac{31}{8}$ on a side

$A = \left(\frac{31}{8}\right)^2 (2 + 2\sqrt{2})$

$A \cong 72.5$ square feet

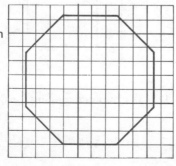

1. a) $\left(\dfrac{P}{4}\right)^2$

    b) $\dfrac{P}{4}$ by $\dfrac{2P}{4}$ Maximum area will have maximum double area, so reflect pasture outline about a line and get twice its area to be a square.

    $$2A = 2(wl) = S^2$$
    $$2wl = (2w)^2$$
    $$wl = w \cdot 2w$$
    $$l = 2w$$

2. a)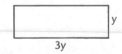

    $x \cdot y = A$ and $x + 2y = \dfrac{cost}{2}$

    So $2A = x \cdot 2y$ is a constant.

    Minimize $x + 2y$ when $x = 2y$ because for a given area, the square has the least perimeter.

    b) $x \cdot y = A$, a constant.

    So $3A$, or $x \cdot 3y$, is also a constant.

    $x + 3y = \dfrac{cost}{2}$.

    Hence, setting $x = 3y$ minimizes cost.

3. 1 · 89
    $\frac{1}{2}$ · 09$\frac{1}{2}$
    $\frac{1}{8}$ · 89$\frac{7}{8}$

    Areas can get smaller without bound.

    Similarly, for any peri-meter, there is no smallest area.

    All perimeters of 180.   Smaller and smaller areas for $P = 180$.

4. $\frac{1}{2}$ · 120
    1 · 60
    5 · 12
    6 · 10

    Perimeters can get greater without bound.

    All areas of 60 square meters

5. a) semicircle   b)   pentagon with all vert-ices on the circle.

    Wall

6. a) Along an arc of a circle passing through the two points.

    b) Along an arc of a circle passing through the two points.

    c) Reflect about the horizontal wall to obtain a figure twice as large that meets the condition of 6a, namely a chain attached to two walls at points equidistant from the corner. For that problem, the solution was to have the chain form an arc of a circle. Under these conditions the top and its reflection below are congruent, and therefore the area of the top is maximized when the area of the entire section is maximized. Hence the chain must follow a circular arc passing through the bolt on the top wall and hooked at the point on the lower wall where the circular arc is perpendicular to the wall.

BIBLIOGRAPHY

Courant, Richard, and Robbins, Herbert. *What Is Mathematics?*
New York: Oxford University Press, 1960, p. 394.

Garfinkel, J., "Exploring Geometric Maxima and Minima."
The Mathematics Teacher (February 1969), pp. 85–90.

Polya, George, *Induction and Analogy in Mathematics. Mathematics and Plausible Reasoning.* Vol. 1. Princeton, N.J.:
Princeton University Press, 1954, pp. 128, 168, 169.

Peterson, John, A., and Hashsisaki, Joseph. *Theory of Arithmetic.* New York: John Wiley & Sons, Inc., 1963, pp. 257–61.

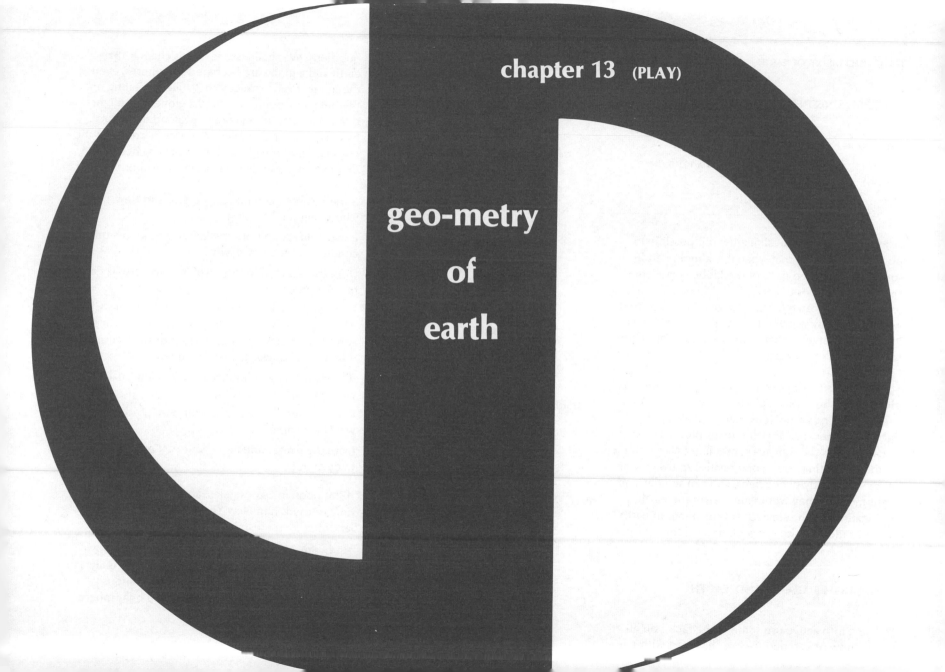

geo-metry

of

earth

MATERIALS NEEDED:

One globe for each student

Mounting for one globe (on campus, for a class period)

Polyurethane and a paintbrush

Using the principles of similarity, it is possible to position a globe so that when it is illuminated by sunlight the lighted parts of the globe correspond to lighted portions of earth. Earth's globe is similar to earth, so we can tell what part of earth is covered with sunlight by watching our globe when similarly aligned, oriented so that its axes are parallel to corresponding axes of earth.

Look at the effect of strong light projected onto a globe. You should see light flooding one hemisphere, creating an obvious shadow line separating light from darkness. This line (considered a great circle) models the sunrise–sunset line on a spinning earth. By rotating your globe relative to the sun or other sufficiently strong light, you can see day and night move as they would move over the earth. Applications of the similarity of sunshine on earth and on a globe is the concentration of this chapter.

13.1 EARTH'S GLOBE AND EARTH

To bring into active use some of the facts you already know or can deduce, we shall start with a few

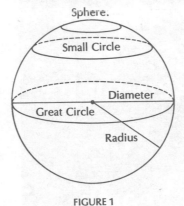

FIGURE 1

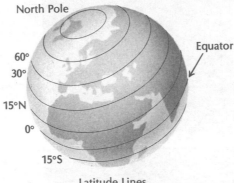

FIGURE 2

questions. We shall assume in this chapter that earth and a globe are both perfect spheres, even though that is not exactly so. While cogitating on the following questions, hold a globe in sunlight or hold it in front of you and pretend that your head is the unmoving sun. Answer according to the orientation of the globe that you have chosen and don't worry about any particular season of the year.

If the sun has been up for two hours in New York, where might it be rising?

In what direction does the light creep across the continent as the day dawns?

Does that mean that the earth is moving to its east or to its west?

If the sun appears directly south of New York City, where does it seem that the sun would be rising? Answer with some names of cities or small countries that would be near the sunrise line.

If the sun appears directly south of New York City, where would it be setting?

Does the sunrise line lie to your east or to your west in daytime?

Does the sunset line lie to your east or to your west in daytime?

If it is noon in Los Angeles, approximately what time of day is it in New York? In Chicago? In Iceland?

These questions may have led you to formulate a sufficiently consistent model of the earth's daily movement to see that the earth is continuously spinning to its east so that the lighted hemisphere

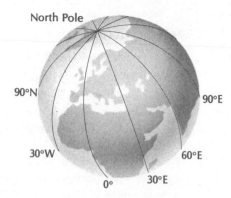

North Pole

90°N

90°E

30°W

60°E

0°

30°E

Longitude Lines

Figure 2 Continued

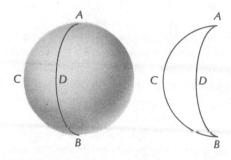

A Piece of the Sphere between Two Longitude
Lines

FIGURE 3

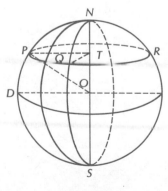

FIGURE 4

appears to be continuously advancing westward. Thus the sunrise line, the line where the lighted portion is advancing, is continuously moving westward, as is the sunset line.

Review of vocabulary connected with a sphere will aid us in communicating. Note the explanatory pictures in Figures 1 & 2 and remember that a man standing on earth is like an extension of a sphere's radius, and that the surface on which he stands is locally like a plane tangent at the radius. Man has used these concepts plus observations of earth's revolutions about the sun to name points and curves on earth.

Notice that all "latitude lines" are small circles save one, the equator. You will also observe that all the "longitude lines" are great circles passing through north and south poles. Using the equator as the zero latitude and the meridian through Greenwich, England as the zero longitude, both latitude and longitude can be expressed as measures of dihedral angles.

EXERCISE 1:

Identify the circle of latitude and the circle of longitude for *P* on the globe in Figure 4. Indicate which angle measures the latitude of *P*. If $\overset{\frown}{NQS}$ is the prime meridian, indicate the angle that measures the longitude of *P*.

EXERCISE 2:

In Figure 5, the point *X* is Miami. Fill in its latitude

and longitude. Miami has a latitude of approximately _____ and a longitude of approximately _____.

EXERCISE 3:

I give you a major city marked Y in Figure 5. Express its position and by checking a good map, see if you can guess its name. In Figure 5, the point Y has a latitude of approximately _____, and a longitude of approximately _____. The city is called _____.

EXERCISE 4:

Explain how you figured out what time it is in Iceland when it is noon in Los Angeles.

Is that time difference correct all year (disregarding local switches to daylight savings time)?

Use what you know of seasons and astronomy to model the earth's travel about the sun. Hold the globe at arm's length and pretend your head is the sun. Your efforts to roughly execute one year's travel will complement the answering of the following exercises.

EXERCISE 5:

Is the number of hours of daylight in a day the same for Copenhagen all year round? Explain why.

EXERCISE 6:

What significance have these three main latitude

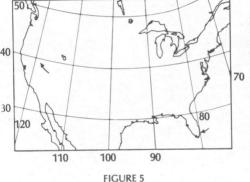

FIGURE 5

lines: the tropic of Cancer, the equator, and the tropic of Capricorn?

EXERCISE 7:

Do the sun's rays reach the north pole every day?

What is the least likely month for the sun to be shining on the north pole?

Is there any physical phenomenon connected with the latitude called the Arctic Circle?

Try making a model of your situation on earth. A globe is similar to the earth, so holding it a certain way will allow us to observe some patterns that are duplicates of those on earth. Get a globe and waterproof it with polyurethane. Erect a sturdy pole anchored well in the ground somewhere where the sun's rays always hit—and where vandalism is low. Then mount a globe at a height so that you can see all parts of it and position it so that your town is hit by sun rays in the same way on the globe as it is on earth. (See Figure 6.)

Correct positioning of your globe on a pole involves two variables that must be taken into account before you can say that your globe is a miniature of earth experiencing light as you do. When oriented properly, the axes of earth and globe will be parallel and the corresponding radii to your town will be parallel. To accomplish this, first orient the globe so that its axis points north and the longitude line through your town is on top; then tip the axis of the globe so that your town is on top. A desirable mounting is one that holds the globe in the proper position and that can be entirely removed from the pole by twisting two wing nuts. Such a thing can be assembled from fittings in a hardware store.

When your globe is positioned properly, light patterns on your globe are similar to those on earth. Much of what is presently happening to earth will be observable or deducible on your globe right now. Because your globe is so placed that it experiences the same shadows as earth, your globe will give you an awareness of which parts of earth are in darkness and which are in light when you are in light, the border between darkness and light denoting sunrise or sunset for the people living in those places.

Sunlight on your mounted globe gives you information, information particularly interesting for certain groups, including buffs on travelling and geography. Those interested in a sister city, sister school or pen pal can experience vicariously some of the life-style of these distant people by such means as watching the pattern of days and nights, and deducing the seasons and temperature.

the sunsets occurring east of you. Three reprints of a map of Mexico compose Figure 7. Please fill in all three with dated sunrise *or* sunset lines, making the first map show that which you observe today or tomorrow and making the other two maps show the same phenomenon at two other dates of the year. Since you probably do not wish to wait 6 months for these dates, you can simulate certain dates by removing your globe from its stand and moving it in relation to the sun. Suggested dates for the second and third maps include March 20, June 21, September 22, and December 21, for on those days the sun is positioned over certain latitude lines. The meaning of each of the dates can be deduced if you are told that on a date we call the summer solstice (June 21), the sun appears to stop moving northward and pauses a day over the tropic of Cancer before returning southward. (See Figure 8).

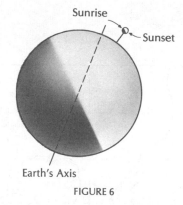

FIGURE 6

EXERCISE 9:

Is the shadow line ever a great circle?

Is the shadow line always a great circle?

Does the shadow line ever go through the poles, making the day and night of the same length?

EXERCISE 8:

Because the mounted globe only registers during the times of day when "the sun is up" (when your town is turned toward the sun), you cannot see both sunrise and sunset creep across any given land. If you live in Florida you can see sunrise creeping across the lands immediately to your west, including the interesting shaped peninsula called Mexico, but you cannot see their sunset on your model. If you live in California, you observe only

EXERCISE 10:

Equinox (equal night) is the time when the length of day equals the length of night.

a)   What occurs when a city experiences an equinox? (Explain in terms of latitude or longitude lines and the shadow line.)

FIGURE 7

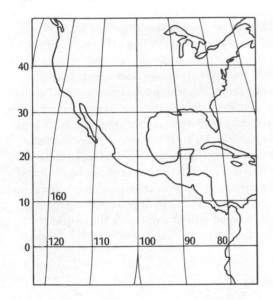

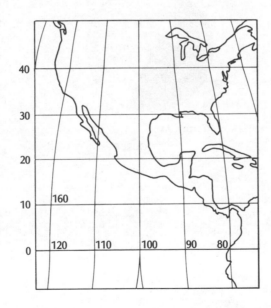

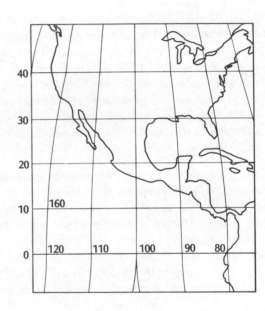

SUMMER SOLTICE

Noon Sun Is Directly Overhead at $23\frac{1}{2}°$ N. Longest Day of Year.

VERNAL EQUINOX

Noon Sun is Directly Overhead at the Equator, on Its Apparent Migration North. Day and Night Are Equal.

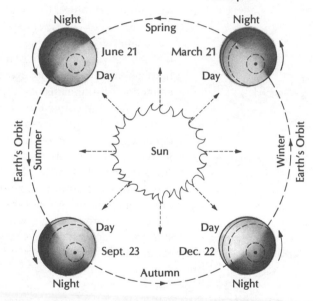

AUTUMNAL EQUINOX

Noon Sun Is Directly Overhead at the Equator, on Its Apparent Migration South. Day and Night Are Equal.

WINTER SOLTICE

Noon Sun Is Directly Overhead at $23\frac{1}{2}°$ S. Shortest Day of the Year.

FIGURE 8

b)   Any two great circles bisect each other. Use this fact to compare the sun's position relative to earth at the time of equinox for people on the equator and those in Chicago and London.

c)   How many times a year do the people of a city like London have an equinox? Position a thin strip of masking tape on your globe to show the shadow line appearing when it is noon in London on the day of an equinox.

EXERCISE 11:

At any given time of year, all people on the same latitude line have the same length of days and therefore the same seasons.

a)   Consider someone living on the tropic of Capricorn. Where else on earth do people have the same patterns of lengths of days but not at the same time of year?

b)   What is the latitude of a locale having the same season in June that Los Angeles has in December? Also state the latitude of Los Angeles.

c)   Compare the length of day, seasons and temperatures of a town at 58° N and a town at 15° N if you know only the latitudes of the two towns.

EXERCISE 12:

Look at your mounted globe today when you haven't looked at a clock for a while. Try to calculate the time where you live by examining the shadow lines. Describe what information you used, and how you deduced the standard time in your locale.

EXERCISE 13:

State the town in which you live, the date, and local standard time.

a)   Where is it now noon, judging by the shadow you see on your mounted globe?

b)   Where is it noon when you see the sunset line approaching your town?

c)   What time is it in Greenwich, England when you see the sunset line approaching your town?

The time in Greenwich, England, is the time reported in many communications. Man has agreed to make all calculations of east and west and all calculations of time from the longitude line passing through Greenwich, England. This line is thus called the *prime meridian*. When time might involve several locales, communicating in Greenwich time eradicates confusion and reduces mental gymnastics.

Astronomers and navigators of all kinds refer to times as if they were in Greenwich. Communication is greatly aided in this way. Pilots would be bothered if, in communicating their plans to different control towers, they had to state times of landings and changes of flight plan in local vernacular. Instead they give all times in terms of the time in Greenwich, England. "Takeoff at 1430 Zula" spoken to a tower near Denver means takeoff is 1430 minus 7 hours, or 7:30 A.M. Denver time. All people in these professions relate to Greenwich time except when reporting events to the local people.

FIGURE 9

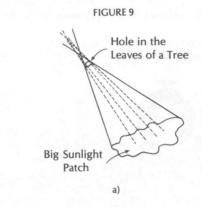

Hole in the
Leaves of a Tree

Big Sunlight
Patch

a)

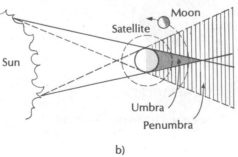

Sun

Satellite

Moon

Umbra

Penumbra

b)

Perhaps you have noted that the cone of light from the sun has a surprising effect when shining through a hole in the leaves of a tree. The sunlight patch on the ground is much bigger than the hole through which the sun shines. (See Figure 9a for a drawing that may trigger your memory.) There is a logical explanation for this difference—that is, the difference between the patch of light created when a flashlight beam passes through a hole and when a collection of sunbeams pass through a hole. Because the sun is so big compared to the flashlight, our sunlight can be considered as passing in a solid circular cone of light whose apex is near the hole in the leaves. In Figure 9a you can see a bit of the sun part and all of the local part of the cone pictured when the outermost light beams cross before entering the hole. (A circular cone remember, is the surface formed by a point and a line through this point tracing a circle, so it has two halves, two nappes.)

Since sun beams converging on earth outline a wide cone, the shadow formed behind any object—including our planet—is conical rather than cylindrical. (See Figure 9b.) This dark conical shadow is called the *umbra*. Because earth is comparatively small, some light rays cross and pass earth to form behind earth a large partial shadow called the *penumbra*. (The lightly shaded area in Figure 9b is the penumbra.) A man aboard a satellite orbiting earth passes through four zones of light, and so he sees different scenes when looking in the general direction of the sun.

EXERCISE 14:

Figure 9b shows the path of a satellite orbiting earth. Make five drawings of the sun as seen by a man in this satellite, labelling your drawings with the time of viewing. His orbit is of four hours duration, and he is closest to the sun at 1800 Greenwich time.

Both the man-made satellite and the moon pass through penumbra, umbra, and back into penumbra as they pass by the dark side of the earth. A man aboard a satellite enters complete darkness followed by a vision of a partially eclipsed sun. Similarly, each place on earth facing the sun sees a different portion of the sun at a time of eclipse.

EXERCISE 15:

Use what you know about umbra to point out where people standing on earth will see a partial eclipse of the sun, and where they will see a total eclipse of the sun. In Figure 10 draw an arrow from each spot on earth to a sketch of what a person living there is expected to see. Draw enough sketches to cover all cases for people who are on that part of earth facing the sun.

The angles of the sun in the last few figures should make you aware of a slight inaccuracy in assuming that the shadow line on earth is a great circle. We have been working on the assumption that the shadow portion of earth was exactly half its surface.

FIGURE 10

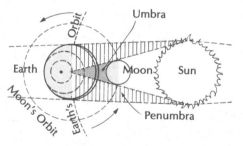

In reality a bit more than half the earth is in sunlight at any given time.

REASONS

1. The sun is so big compared to earth that the light extends beyond the diameter to the points where the outermost light beams are tangent to the earth. (See the unbroken lines of Figure 9b)

2. The light is diffused or spread by the atmosphere.

3. Diffraction bends light inward toward the object—a curved path differing slightly from the previous tangential path.

    Of these three reasons, the first is the major reason for any discrepency you might have noticed—and it is the one we wish to emphasize because it is the only one that is purely geometric.

EXERCISE 16:

Have you ever noticed that the shadow of a leaf was very fuzzy if the leaf was more than a few inches from its shadow? Why is that so?

There! Now you see that geometrical knowledge is of use in even more places than you suspected in your most liberal moments. The similarity of earth and globe make shadow clocks and home learning devices possible. Other geometric effects of sunlight on earth included the umbra and its relevance to space travel and eclipses.

ANSWERS TO EXERCISES

1. Latitude circle for P is $\overline{RPQ}$.

   The latitude of $P = m(<DOP)$ north.

   Longitude circle for P is $\overline{RPDS}$.

   The longitude of $P = m(<QTP)$ west.

   We say P is so many degrees north and so many degrees west.

2. 26° north; 80° west.

3. 38° north; 119° west; Reno, Nevada.

4. Since in 24 hours the earth turns 360° rotation, it turns 15° per hour. The first belt is centered in Greenwich, England around the 0° meridian. Each state or nation decides its time according to which 15° meridian they are nearest.

   Los Angeles is near longitude 118° and Iceland is near longitude 18°

   Los Angeles is on the time centered at 120° and Iceland is on the time centered at 15°.

   $120° - 15° = 105°$

   $\dfrac{105°}{360°} = \dfrac{X}{24 \text{ hours}}$

   $X = 7 \text{ hours}$

   Iceland is 7 hours later, or 7 : 00 P.M. when it is noon in Los Angeles. The time difference is the same all year because the earth rotates at the same speed all year.

5. No. Earth receives sun's direct rays at different latitudes as it rotates around the sun (see Figure 8); in December, Antarctica gets more sun than the Arctic and Copenhagen.

6. On December 21, the sun appears directly over the tropic of Capricorn. This is the furthest south that it appears overhead, so we in northern countries experience the shortest day of the year. The sun appears to then travel northward. Astrologers and nature lovers consider the beginning of the year to be March 20th when the sun is positioned over the equator. The sun over the tropic of Cancer and back over the equator mark the beginning of northern hemisphere's summer and fall seasons, respectively.

7. No, in December, the sun is near the tropic of Capricorn. The Arctic Circle is the furthest north the sun's rays can reach on December 21. Similarly, the Antarctic Circle is designated as the circle beyond which the sun's rays cannot reach around June 21.

8. The map of Mexico will show shadow lines at various angles, ranging from NW–SE to NE–SW, depending on the times of year you choose.

9. Always a great circle. The shadow lines go through the poles when the day and night are of the same length, twice a year when the sun is over the equator.

10. a) An equinox occurs in London whenever the latitude circle of London is bisected by a shadow line. Since all latitude lines are parallel, this incident is simultaneous with the bisection of the equator and southern latitude lines. Thus the shadow line goes through the poles, and the sun is over the equator.

    b) The equator is always bisected by a shadow line. Those living on the equator always experience equinox. When the latitude line of Chicago is bisected by the shadow line, so is that of London because the sun must be over the equator.

    c) Twice a year, on March 20 and September 22, the sun is over the equator. On those dates, people all over earth experience an equinox.

11. a) Tropic of Cancer.

    b) Los Angeles has latitude of approximately N 35°. A

town near S 35° latitude has the same season at an opposite time of year.

c)   At 15° N, more long days and fairly constant warm weather; May and July are likely to be the hottest months having the longest days. At 58° N, we have an exceptionally short day around December 21 (6 hours of sunlight) and an exceptionally long day around June 21, when the weather is warmest.

14.

| 1800 | 1930 | 2000 | 2030 | 2100 |

15.

16.   The umbra is so small a cone relative to the distance from the leaf to the ground that you see partial shadows of the penumbra on the ground, and sometimes no umbra at all.

BIBLIOGRAPHY

Hutchinson, M. W. *Geometry: An Intuitive Approach*. Columbus, Ohio: Charles E. Merrill Publishing Co., 1972, pp. 294–6.

Minnaert, Marcel Gilles Josef. *The Nature of Light and Colour in the Open Air*. Translated by H. M. Kremer-Priest. Revised by K. E. Brian Jay. New York: Dover Publications, Inc., 1954.

Schaaf, William L. *Basic Concepts of Elementary Mathematics*. New York: John Wiley & Sons, Inc., 1966, p. 318.

School Mathematics Study Group. *Studies in Mathematics*. Vol. 9. Revised Edition. Palo Alto, Calif.: Leland Stanford Jr. University, 1963. pp. 350–3.

Smart, James R. *Introductory Geometry, An Informal Approach*. Monterey, Calif.: Brooks/Cole Publishing Co., 1969, pp. 162–4.

Smith, Bernice. "Sunpaths That Lead to Understanding." *The Arithmetic Teacher* (December 1967), pp. 674–7.

Sullivan, John J. "Problem Solving Using the Sphere." *The Arithmetic Teacher* (January 1969), pp. 29–32.

chapter 14

volume

This chapter is devoted to the measure of solids. The measure of a solid is called volume. Chapter 11, devoted to the study of area, demonstrates the usefulness of the square shape both in tessellating and in further subdividing into like shapes. Analogous to this choice, we denote the solid cube as our standard of volume measure and assign the volume of 1 cubic unit to a cube whose edge measures 1 unit in length.

The development in Chapter 11 progressed from approximations of area via grids and weighing to exact calculations aided by reasoning. Our development of volume will be similar in that we start with physical experimentation in approximating volume and progress to calculations aided by logic. In addition, we will use what we study in this chapter to conclude with a more sophisticated estimation of volume, the summing process underlying integral calculus. All shortcuts to figuring volume are motivated by common knowledge and experience with well-known objects.

Dice and children's building blocks are convenient cubes with which to estimate volume of such solids as cones and hemispheres. Approximation is easier when accompanied by specific questions such as these relating to a cone: "Guess how many of the little cubes will fit into this cone? How many big cubes? What size cubes would you use if you were trying to guess the exact volume?" For more accurate estimating, each cube can be sectioned so that it is made up of smaller cubes, each similar to the original. How many cubes of what dimension can you manufacture from a cube with edges 1 centimeter long? You can get 1000 cubes of $\frac{1}{10}$ centimeter on an edge and you can get 8 cubes of $\frac{1}{2}$ centimeter on an edge.

Estimates of volume are very instructional, particularly when arrived at by comparison with other solids of known volume. The hardest to estimate visually is the volume of a sphere; but even that can be demonstrated to be less than $(2r)^3$, by comparison to a cube. The sphere's volume is obviously less than that of the smallest cube containing it. Seeing the sphere placed in this cube allows a better estimate of comparative volumes; the sphere appears to take up about half of the total volume within the cube, and thus has volume nearer to $4r^3$. (Actually, volume inside a sphere has been proven to be $4/3\pi r^3$, only a shade more than the $4r^3$ estimated physically.) The sphere's exact volume is hard to guess from physical examination, but guessing followed by experimental checking in filling things with sand or water leads to some other relationships.

14.1 VOLUME ENCLOSED BY FAMOUS SURFACES

Sometimes our efforts at calculating volume by comparison astound us by yielding exact volumes. Here is one such comparison that adapts well to experimentation. Let us explore this relationship via an instructional demonstration using sand (or water), a right prism and a pyramid built on the same base and having the same height.

Fill the model of the pyramid with sand. Empty the sand from the pyramid of square base into the square prism. Repeatedly fill the pyramid and pour its contents into the prism, keeping count of how many pyramids fill the prism. (See Figure 3) Because it takes three pyramids full of sand to fill the prism, we say that the volume of that pyramid is $\frac{1}{3}$ the volume of its related prism.

$$V_{\text{right square pyramid}} = \frac{1}{3} V_{\text{right square prism}} = \frac{1}{3}\left(A_{\text{base}} \cdot h\right)$$

Will it be true no matter what size square base is chosen? Will it be true for other than square prisms and their related square pyramids? It is particularly easy to check the relationship between prisms and pyramids of circular base (called cylinders and cones). The world is overflowing with cylindrical tin cans and anyone can easily roll a sheet of paper into a cone of the same circular opening and make a sand test.

Since the volume of each pyramid appears to be $\frac{1}{3}$ that of the related prism, the inquisitive might like to experiment with three congruent pyramid models to see if he can fit them together, whole or in pieces, to make a prism. There are commercial multicolored plastic models of triangular pyramids that slip whole, like puzzle pieces, into the open end of a triangular prism. It can be proved mathe-

FIGURE 1

14.1 VOLUME ENCLOSED BY FAMOUS SURFACES ① 191

FIGURE 2

FIGURE 3

FIGURE 4

a)   b)   c)

matically that, no matter the shape of the base,

$$V_{\text{right pyramid}} = \frac{1}{3}\left(A_{\text{base}} \cdot h\right).$$

We needed discussion and some equipment in order to really believe the relationship between a right pyramid and its related prism. Some other relationships between solids are sufficiently obvious that they can be explored with a little equipment mixed with some logical reasoning. Brief examination of a rectangular prismoid solid with dimensions like those pictured in Figure 4a should yield its volume, for it is two cubes placed side-by-side. The volume of the solid in Figure 4b can be visualized by counting two layers of the cube in 4a. Even more interesting is the use of common sense to see that the triangular solid in Figure 4c is half as big as a rectangular solid whose base is 1 by 2.

The relationships of Figure 4 all follow a pattern, the putting together of like things to form a prism. We notice also that the relation between the prisms of 4a and 4b suggests that the volume of a solid prism is numerically the area of the base times the number of cubes layered in height. We can express this simply as $V = A \cdot h$, where A represents the area of the base and h represents the height. Will this also be true for a right circular cylinder and a right cylinder having an irregular base?

As a deck of cards has the same volume no matter how they are stacked, so a prism that *slants* will have the same volume as the corresponding right prism. A *right prism* is a prism in which the lateral edges are perpendicular to the base. An *oblique prism* is one that is not a right prism. You can experimentally verify, with cards, wood chips or table coasters, that the volume within an oblique prism is the same as that within a right prism having the same base and height.

Imagine an ordinary deck of cards alongside a deck of cards of the same thickness but of a circular shape. If the area of the face of a circular card is the same as the area of the face of a rectangular card, will not the volume of the 2 decks be the same? Surely they will! The 2 decks are each composed of 52 cards of equal volume. This comparison between cylindrical solids does not rest upon the shape of the faces being compared (nothing special about rectangle and circle). If the faces are of equal area and the heights are equal, then the volumes are equal, even for non-prisms. This is the reasoning behind a general principle first verbalized by an Italian mathematician named Cavalieri.

Cavalieri's principle:
If two solids are included between the same two parallel planes, and if each plane parallel to these planes cuts cross-sections of equal area in the solids, then the volumes of the two solids are equal.

This principle applies alike to pyramids, cones, prisms, cylinders and weird shapes. Note the illustration of the pyramids and cone in Figure 6. If C_1, C_2, and C_3 have equal area (and that is true for any place P_1 is placed, including the base), then the volumes of the three solids are the same.

FIGURE 5

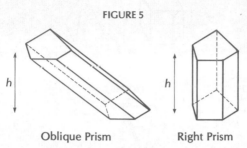

Oblique Prism   Right Prism

FIGURE 6

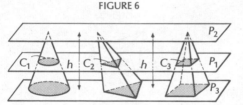

Three Solids with Equal Cross Section and Equal Volume

FIGURE 7

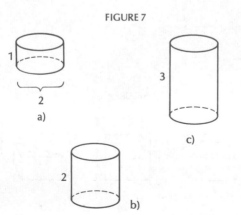

Given a square pyramid such as that used in the sand demonstration, we can visualize a cone like that in Figure 6 having the same height and having cross-sections of area equal to those of the square pyramid. By Cavalieri's principle, the volumes of the pyramid and cone are the same; hence, $V = \frac{1}{3} A \cdot h$ is also true for the cone. By the same reasoning, the volumes are the same for all pyramids and cones having the same height and bases of equal area; so $V = \frac{1}{3} A_{base} \cdot h$ is a formula applicable to all pyramids and cones.

Do any of the experimental and logical particulars discussed so far lead to other geometrical facts? Yes, knowing how to find the volumes of the solids pictured in Figure 4 gives you good basis for finding the volumes of the solids pictured in Figure 7: Figure 7a represents a solid prism 1 unit high and resting on base of area π. Its volume is π cubic units, so it follows that by adding layers we get 2π and 3π cubic units as the volumes of Figures 7b and 7c. This idea will also carry through for cylinders of irrational height and radius.

For any prism, $V = A_{base} \cdot h$

14.2 INTRODUCING INTEGRATION METHODS FOR IRREGULARS*

In finding the volume of the cylinders in Figure 7, we were using the integration (summing) principle of calculus. We summed circular prisms 1 unit high. The idea of integration is invaluable for finding volumes of those solids whose cross-sections are not congruent. Summing is basic to the integration method of calculus. I will demonstrate the basics of this method by approximating the volume contained in a cone 5 units high that has a circular base of radius 5 units.

We use the cone as an example because it is an interesting symmetrical shape whose successive approximate volumes can be compared to its exact volume, $\frac{1}{3} A_{\text{base}} \cdot h$ (the formula we derived in the preceding section). Before you study the summing method, you might make a quick calculation of this cone's exact volume so that you can see how close we come with each approximation.

If we approximate the volume of the cone in Figure 8a with small cylinders of equal height and lessening radius (see Figure 8b) we get for our first approximation the sum of the volumes of the five small cylinders.

$$V = \pi \cdot 5^2 \cdot 1 + \pi \cdot 4^2 \cdot 1 + \pi \cdot 3^2 \cdot 1 + \pi \cdot 2^2 \cdot 1 + \pi \cdot 1^2 \cdot 1$$
$$= \pi (25 + 16 + 9 + 4 + 1) = 55\pi$$
$$V_{\text{cone}} < 55\pi.$$

You can see in Figure 8c how much better an approximation we get in taking 25 small cylinders of height $\frac{1}{5}$ unit.

FIGURE 8

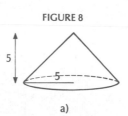

a)

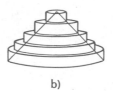

b)

c)

$$V = \frac{1}{5} \cdot \pi \cdot \left[5^2 + \left(\frac{24}{5}\right)^2 + \left(\frac{23}{5}\right)^2 + \ldots + \left(\frac{2}{5}\right)^2 + \left(\frac{1}{5}\right)^2 \right]$$
$$= \frac{\pi}{5} \left[\frac{25 \cdot 26 \cdot 51}{25 \cdot 6} \right] = \frac{221}{5} \pi$$
$$V_{\text{cone}} < \frac{221}{5} \pi$$

Although 25 cylinders provides a better estimate than 5, our cylinders are still enclosing the cone. To have an idea of how accurate this estimate is, we should consider the volume of 25 slender cylinders enclosed by the cone. We know that the volume of the cone is between the above result and that of summing 25 cylinders inside the cone. (Verify for yourself that the slender inner cylinders will range in radius from $\frac{24}{5}$ to 0, making the inside sum that of the outer minus the volume of $\frac{1}{5} \cdot \pi \cdot 5^2$). Using 25 cylinders gives only sufficient accuracy to guarantee that your estimate is within 5π of the real volume.

$$\frac{196\pi}{5} < V_{\text{cone}} < \frac{221\pi}{5}$$

Ideally, we would like to have more pieces—smaller ones with which to approximate the volume of the cone. We can get as accurate an estimate as we wish, but the finding of exact volume by summing awaits the tool of calculus. The integral of calculus essentially sums an infinite number of infinitely thin pieces. Integration is a method of quickly finding that number which is the limit of the inside volumes as an increasingly greater number of pieces are used. *It is provable*, using calculus, that the actual volume of the cone in Figure 8a is $\frac{1}{3}\pi \cdot 5^3 = 41\frac{2}{3}\pi$ cubic units.

The volume of any weirdly-shaped solid can be found by adding together slices between planes—slices whose volume is known or can be estimated using the preceding method. The summing of the volumes you find for these pieces is just a wider application of the integration principle. The practice of summing volumes is a sound way of getting volume estimates of many irregular solids we deal with in physical problems. This practice also provides a firm base on which the concepts of calculus can be built at a future time. (The use of the full tool of integral calculus is possible when the boundaries of the solid can be expressed algebraically. An understanding of Cartesian coordinates, Chapter 16, is needed so that one can express boundaries algebraically and thus take advantage of calculus and other fields.)

14.3 EXERCISES FOR CHAPTER 14

Leave the symbol, π, in the answers.

I. 1. Find the volume of each solid:
    a) Tin can 10 centimeters high having a circular end of radius 8 centimeters.
    b) TV box that is 2 feet long on each edge.
    c) Stereo cabinet and its legs, in cubic inches. The cabinet is 2 feet wide and 1 yard long and 33 inches high after its legs are removed. The 4 legs are cylindrically shaped, 4 inches long and 1 inch in diameter.
    d) An ottoman $1\frac{1}{7}$ feet high, whose top and bottom are regular hexagons of circumference 60 inches.

2. Archimedes is showing his king the principle behind his "Eureka discovery." He drops a gaudy bracelet into a cylindrical vessel 4 inches in diameter partially filled with water and the water rises $\frac{1}{4}$ inch. What is the volume of the gaudy bracelet?

3. a) What is the capacity of a freestanding cylindrical firebowl (on the order of a topless tuna can)? The fireplace is of stone, laid 5 inches thick throughout, and reaching to a height of 25 inches above the floor boards. The outside diameter is 3 feet.
    b) What volume of stone is needed to construct the fireplace?

4. Calculate the number of board feet in 92 studs, 2 inches by 4 inches and 8 feet long. (*Board feet*: the number of feet of 1 by 12 lumber that is equivalent in volume.)

5. How much sand is there in a conical sandpile 44 feet in circumference at the bottom and 5 feet high? (Express in cubic feet, estimating π as $\frac{22}{7}$.)

6. Find the volume of the solid generated:

a) when a right triangle of base *b* and altitude *h* revolves about its side *h*.

b) when the triangle revolves about its side *b* (see Figure 9).

c) Would you expect these two volumes to be equivalent? Under what conditions would they be equivalent?

7. A modification of any solid can be accomplished by slicing the solid horizontally into two solids. When a cylindrical solid is so sliced, both pieces are cylindrical. Yet, when a conical solid is sliced this way, only one piece is similar to the original. The piece shaped differently is called a *frustum* of the solid. Man-made objects frequently model frustums of cones and pyramids. Lampshades and wastepaper cans usually outline frustums of cones; the pedestal of a statue or trophy is usually a frustum of a pyramid.

a) How much wallpaper is needed to cover the outside surface of the wastebasket in Figure 11? What is the capacity of the wastebasket if we consider it of negligible thickness?

The volume of a conic frustum is that remaining after the volume of the little cone sliced off has been subtracted from that of the whole cone.

b) Find the approximate volume of the

FIGURE 9

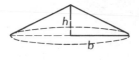

FIGURE 10
Frustums of Cones

FIGURE 11

10
18″
6

FIGURE 12

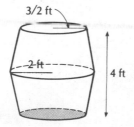

3/2 ft
2 ft
4 ft

whiskey barrel pictured in Figure 12. It is 4 feet high and shaped like two congruent conic frustums placed with their wide ends together. The barrel has a radius of $\frac{3}{2}$ feet at the top and 2 feet at the center.

c) What volume of water has yet to be added to fill a conic container already filled 2 feet deep? The conic container is 5 feet deep and has a radius of $\frac{5}{2}$ feet at the top.

8. a) Find the volume within a solid prism having equilateral triangles as bases, when the triangles measure $2\sqrt{3}$ centimeters on a side and the height of the prism is 3 centimeters.

b) Find the volume within a tetrahedron built on the same triangular base.

c) Find the volume of water needed to fill the bottom 1 centimeter of the tetrahedron described in Exercise 8*b*. (Hint: Use previous results.)

II. True or False

9. Given that a right prism and an oblique prism have the same volume and the same base, the right prism has the smaller surface.

10. Given that a right pyramid and an oblique pyramid have the same volume and the

same base, the right pyramid has the smaller surface.

III. In each of the following, two solids will be described. If V is the volume of the first solid, fill in the blank with the volume of the second solid.

11. A circular prism having radius of base 3 inches and a height of 6 inches has a volume V. What is the volume of a cylinder:

a) with base of twice the diameter?

b) with twice the height?

c) with twice the area of the base?

d) with both the height and diameter doubled?

12. Repeat the previous problem for a cone of volume V. Will changes in height and diameter have the same effect upon a cone as they have upon a circular cylinder?

13. A well-proportioned man is of volume V. What is the approximate volume:

a) of a man five times as big and similarly proportioned?

b) of a man stretched twice as tall, but with hips, shoulders etc. of the same dimension?

14. A 1-inch piece of bone has volume V.

What is the volume of that length of bone:

a) if the radius of the bone is doubled?

b) if the diameter of the bone is tripled?

15. Explore spatial characteristics of a tropical spider reputed to be seven times the size of the black widow spider, but similar in build:

a) If V is the volume of the black widow, what is the volume of the tropical spider?

b) If W is the weight of the black widow, what is the approximate weight of the tropical spider?

c) If A is the area of a cross-section of the black widow's leg, what is the area of the corresponding cross-section of the leg of the tropical spider?

d) If L is the weight borne by the black widow's leg, what is the weight borne by the leg of the tropical spider? (Assume here that the two spiders have the same overall structure.)

e) The leg cartilage in the black widow is sufficient to bear its weight. If the same type of material is found in the leg of the tropical spider, what must be true about the supporting ability of that type of cartilage?

1. Discuss the relation of man's size and shape to the ability of the neck-bone to support weight. Use helpful examples from the animal kingdom such as a chubby person, a skinny person, a giraffe, and a rhinoceros.

2. Discuss the relation of man's size and shape to heat maintenance and loss. Use chubby and skinny people and other specific examples from the animal kingdom, if helpful.

3. A parallelogram can be sliced vertically and refit to form a rectangle. This visually corresponds to a proof that the area of a parallelogram is the same as that of a rectangle having the same base and height.

    Can you do something similar with an oblique hexagonal prism, i.e. can you slice it and refit it into a right hexagonal prism?

    Can you do it with an oblique rectangular prism?

    State your opinion on slicing of solids as a means of calculating volume.

4. Post Office Problem: Find the shape of maximum volume for a box that must comply with the U.S. postal regulation that girth (distance round) plus length is at most X units.

5. We know how to find the area and volume of any circular cone. By rewriting algebraically, we can see that the volume inside a cone of a given surface is related to the product of base and lateral surface areas.

$$A = \pi r^2 + \pi rs$$

$$= 2\pi r^2 + \pi r(s - r)$$

$$V = \frac{\pi r^2 h}{3}$$

$$V^2 = \left(\frac{\pi}{3}\right)^2 r^4 h^2$$

$$= \left(\frac{\pi}{3}\right)^2 r^4 (s^2 - r^2)$$

$$= \left(\frac{\pi^2}{9}\right) r^4 (s - r) \cdot (s + r)$$

$$= \frac{\pi}{9} r^2 r(s - r) A$$

$$= \frac{\pi}{18} A \, 2r^2 \cdot r(s - r)$$

$$= \frac{A}{18\pi} \cdot 2\pi r^2 \cdot \left[\pi r(s - r)\right]$$

If we wish to maximize volume we will also be maximizing volume squared. For a given surface area; V^2 is maximized as the product, $(2\pi r^2) \cdot r(s - r)$ is maximized. Let us consider all cones of surface area, say, 40 square units. Fill in more rows of the table below in an attempt to find a distribution of the 40 square units to maximize volume. Obviously, your findings on best shape of cone can be generalized to other sizes of cones other than 40 square units.

(Base) πr^2	$2\pi r^2$	(Lateral) $\pi r(s-r)$	$2\pi r^2 \cdot \pi r(s-r)$	(Volume^2) $\dfrac{A}{18\pi} \cdot 2\pi r^2 \cdot \pi r(s-r)$
10	20	30	600	$\dfrac{40}{18\pi} \cdot 600$
5		35		

What proportion of the total should be the base, if volume is to be maximized?

I. 1. a) $640\,\pi$ cubic centimeters

b) 8 cubic feet

c) $(28512 + 4\pi)$ cubic inches

d) $2700\sqrt{3}$ cubic inches

2.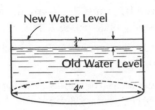

New Water Level

Old Water Level

$\frac{1}{4}''$

$4''$

Bracelet has volume of π cubic inches

3. a)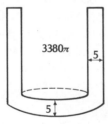

3380π

5

5

b) 4720π cubic inches of stone needed.

4. $490\frac{2}{3}$ board feet

5. Approximately $256\frac{2}{3}$ cubic feet

6. a) $\frac{1}{3}\pi b^2 \cdot h$ cubic units

b) $\frac{1}{3}\pi h^2 \cdot b$ cubic units

c) Not generally the same, but they will be the same when the right triangle is isosceles (when $h = b$).

7. a) The height of the whole cone is 45 inches.

$S = (295.2 - 100)\pi$ square inches.

$V = 1176\pi$ cubic inches.

b) One half barrel has volume $\frac{37}{6}\pi$ cubic feet.

The barrel has volume $\frac{37}{3}\pi$ cubic feet.

c) Frustrum of the cone $= \frac{\pi}{3}\left(\left(\frac{5}{2}\right)^2 \cdot 5 - (1)^2 \cdot 2\right) = \frac{39}{4}\pi$

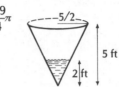

5/2

5 ft

2 ft

8. a) $9\sqrt{3}$ cubic centimeters

b) $3\sqrt{3}$ cubic centimeters

c) The top of the water will fill a triangle whose dimensions are $\frac{2}{3}$ as big as those of the base to which it is similar. So, the area of the top of the water is $\frac{4}{9}$ of $3\sqrt{3} = \frac{4}{3}\sqrt{3}$ square centimeters.

Frustum: $3\sqrt{3}$ cu. cm. $- \frac{1}{3}(\frac{4}{3}\sqrt{3})$ (2) cu. cm. $= \frac{19}{9}\sqrt{3}$ cu. cm.

II. 9. True, because the surface area is dependent upon the slant height rather than the altitude of the solid.

10. True.

III. 11. a) Twice the diameter $\longrightarrow$ twice the radius $\longrightarrow V = \pi 6^2 \cdot 6 = 4V$

b) $2V$

c) $2V$

d) $8V$

12. $V_{cone} = \frac{1}{3} V_{cylinder} \longrightarrow$ same effects from changes.

13. a) $125V$ b) $2V$

14. a) $4V$ b) $9V$
15. a) $343V$
     b) $343W$
     c) $49A$
     d) $7L$
     e) It must be able to bear seven times the weight in each unit of cross-section.

ANSWERS TO PROBLEMS

3. Yes for any rectangular prism because opposite faces are congruent. No for the general hexagonal prism.

4.
$$2w + 2h + l = X$$
$w \cdot h \cdot l$ maximum when
$4w \cdot h \cdot l$ maximum when
$2w \cdot 2h \cdot l$ maximum.

A cube has the maximum volume because its three dimensions are of equal size.

So $2w = 2h = l$ will make $22 \cdot 2h \cdot l$ maximum.

Description of box: It is a square prism, $w = h$ and $l = \frac{1}{2}$girth $= 2w$.

5. We get the maximum volume within a cone of 40 square units when the area of the base is $\frac{1}{4}$ the total surface area.

Is this true for any cone, i.e., is base $= \frac{1}{4}$ total surface always the proportion giving maximum volume?

BIBLIOGRAPHY

Archimedes. *The Method.* Edited by T. L. Heath. New York: Dover Publications, Inc., 1912.

Boltyanskii, V. G. *Equivalent and Equidecomposable Figures.* Indianapolis: D. C. Heath & Co., 1963, pp. 37–66.

Haldane, J. B. S. "On Being the Right Size." *The World of Mathematics.* Vol. 2. Edited by James R. Newman. New York: Simon & Schuster, Inc., 1956, pp. 952–7.

Horn, Sylvia. *Learning About Measurement.* Pasadena: Franklin Publications, Inc., 1968, pp. 54–59.

National Council of Teachers of Mathematics. *Measurement, Topics in Mathematics for Elementary School Teachers.* Booklet 15, 1968.

Polya, George. *Induction and Analogy in Mathematics. Mathematics and Plausible Reasoning.* Vol. 1. Princeton, N.J.: Princeton University Press, 1954, pp. 128–31, 137–41.

Smart, James R. *Introductory Geometry, An Informal Approach.* Belmont, Calif.: Brooks/Cole Publishing Co., 1969, pp. 165–75.

chapter 15

area
&
volume

MATERIALS NEEDED:

Two pieces of glass and a soap-bubble blown (optional)

In Chapter 12 you noticed that some polygons enclose more area than others of the same peri-meter. Analogously, the volumes enclosed by dif-ferent polyhedrons of the same area differ. Each type of polyhedron has proportions yielding maxi-mum volume. For a given surface area, some shapes of boxes have a greater volume than other shapes. Some tetrahedrons contain a greater volume than others of the same surface area. And so on for cylinders and other types of closed surfaces. Each type of closed surface has certain properties, each has a shape enclosing the maximum volume for a given amount of surface. You began your explora-tion of such problems when you worked on the problems of Chapter 14—finding the box of maxi-mum volume meeting post office specifications, the best shape of human for heat maintenance, and the shape of cone that maximizes volume.

Maximizing the volume for a given surface is the same thing as minimizing surface for a given vol-ume, something of major concern to the people in the packaging industry. These are two wordings of the same problem, since they are mathematically equivalent. One approach to a problem can be dif-ficult while the equivalent wording leads to an easy solution.

15.1 MAXIMIZING VOLUME, MINIMIZING SURFACE

We will approach most volume-surface relation-ships by considering maximizing of volume enclosed by a certain type of surface having a specified area. Let us begin by exploring the relationship between volume and surface of boxes.

EXERCISE 1:

a)   Which uses the least amount of cardboard: A box of dimensions 4 by 4 by 10 units, or a box of 4 by 5 by 8 units?

b)   Which of the boxes encloses a greater volume for the amount of cardboard used?

State the ratio $\dfrac{V}{S}$ for both boxes.

EXERCISE 2:

Now let's turn the problem around and examine boxes of a given surface area, S. Consider an S of 96 square units and find that box (right rectangular prism) having the greatest volume. Using symbols, x, y, and z for lengths of the three edges, fill in the table below with all the different boxes you visual-ize in the course of your exploration, and then circle the one describing the box of maximum volume.

x	y	z	Volume xyz
2	8	2	32

In exercise 1, you found that a box of 4 by 5 by 8 has less surface area than the box of equal volume with a shape of 4 by 4 by 10. Is there a shape of box with a given surface area containing maximum volume? (Or if you wish, reword it, "For volume being constant, is there a rectangular prism of minimum surface area?")

EXERCISE 3:

Consider a right rectangular prism having edges of length x, y, and z. The following questions are to help you see a general relationship between surface area, S, and volume. Then you can see a way to express maximum volume for a given S.

a) What is another name for $xy + yz + xz$?

b) State another name for $(xy) \cdot (yz) \cdot (xz)$.

c) Will maximizing $(xy) \cdot (yz) \cdot (xz)$ maximize volume?

The development of a relationship between V and S in the box of maximum volume involves only a little algebra and use of what we learned about

the square containing the maximum area $(A_{square} > lw$ wherever $l \neq w)$.

Given that we have a box of dimensions x, y, and z, $S = 2xy + 2yz + 2xz$.

$S/2 = xy + yz + xz$ and $V^2 = xy \cdot yz \cdot xz$.

So V^2 maximum when $[xy \cdot yz] \cdot xz$ maximum.

This necessitates, for xz fixed, that $[xy \cdot yz]$ maximum.

So $xy = yz$ when V^2 maximized while xz is held fixed.

V^2 can also be written $xy[yx \cdot xz]$.

So by the same reasoning, $xz = yz$ when V^2 maximized while xy is held fixed.

Hence $xy = yz = xz$

So $x = z$ and $y = x$, which implies that $x = y = z$.

So V^2 can be written x^6.

Also $S/2$ can be written as $3x^2$.

Hence $V = x^3$ and $S = 6x^2$.

Check this relationship against your answers to the questions of Exercises 2 and 3. It should turn out that a cube is, in each case, the box of maximum volume for a given surface; the closer a box is to the shape of the cube, the closer its volume is to maximum.

EXERCISE 4:

You are going to use a wooden plank on which you wish to pin a deep open container for a quantity of bird seed. You wish to use a part of this plank as the bottom of the feed container and just fashion the lateral surface.

If you have a certain rectangle of metal, how would you bend it so that you form the lateral surface of a container having maximum volume? Explain.

EXERCISE 5:

You have a triangular prism having height h and area of base, A. Consider the collection of all prisms and cylinders having the same height and the same area of base. (They all have the same volume.) Is there any shape that has minimal surface area? (Hint: Reread the last exercises of Chapter 11.)

You have ample evidence in your environment that the closed surface containing maximum volume for the area is the sphere. Notice what happens as you blow a bubble of soap or bubble gum; the surface stretching to meet the influx of air forms a spherical shape. This is a physical demonstration, but not a mathematical proof, that the sphere accomodates the maximum volume for a given surface area.

Similarly, the soap bubble, restricted, provides a physical demonstration that the maximum volume under a certain restriction is contained within the cylinder. Consider a soap bubble, blown so that it touches two parallel wetted pieces of glass as in Figure 1. It immediately forms a circular cylinder between the two pieces of glass. For the volume and pressure, the lateral part, the soap film

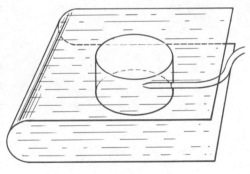

FIGURE 1

part, has minimal surface. This is a physical demonstration, not a proof, of the maximality of the cylinder's volume for a given surface area. It is also what you expected from analogy and from your work in Chapter 12.

We expect, by analogy with Chapter 12, that the sphere and cylinder might enclose maximal volumes. This information is correct and we can use it to work problems, but many other analogies to isoperimetrics in two dimensions are fallacious. Now that we are in 3-space, the number of variables has increased and the complexity has increased to where the calculus of variations is needed to solve some of the problems analogous to the isoperimetrics we met in Chapter 12. Making this study as complete as that in Chapter 12 is therefore impossible. We will instead explore some problems that are essentially applications and extensions. You will find interesting supplemental material in George Polya's questions[1] and in explanations and experiments on page 198 of Young and Bush.[2]

EXERCISE 6:

a) Which encloses a greater volume in proportion to the amount of material used—a cylinder 8 inches

[1] George Polya, *Induction and Analogy in Mathematics, Mathematics and Plausible Reasoning*, vol. 1, (Princeton, N. J.: Princeton University Press, 1954), pp. 128–41, 155–8, 166, 168–89.

[2] Grace A. Bush, and John E. Young, *Geometry for Elementary Teachers* (San Francisco: Holden-Day, Inc., 1971), p. 198.

high and of a radius of 3 inches or a cone of the same dimensions?

b)   Is this true for any cylinder and cone having the same dimensions?

Problem 5 of Chapter 14 asked about the distribution of surface that yields a cone of maximum volume. Whether or not your finding conformed to your expectations, you deduced that a cone has maximum volume when $\frac{1}{4}$ of the total surface area is in the base. You might experiment and calculate to check out a conjecture that this proportion will be maximum for a cylinder too. Remember, that which at first seems likely in three dimensions may not follow analogously from what we know to be true in two dimensions.

EXERCISE 7:

Try to guess the type of polyhedron with 8 faces that, for a given surface area, contains the maximum volume. State a conjecture for the appearance of the polyhedron of n faces containing maximum volume for a given surface area.

As an example of the complicated connection between problems in two and three dimensions, reconsider the conjecture of Exercise 7. You were drawing an analogy if you made this conjecture: "If there is a regular solid with n faces, it yields the maximum volume." George Polya reports on page 188 of *Induction and Analogy in Mathematics*, volume 1 of *Mathematics and Plausible Reasoning* (Princeton, N.J.: Princeton University Press, 1954),

that, plausible though this conjecture may seem, it turns out wrong in two cases out of five.

Conjecture is correct for $n = 4, 6, 12$.

Conjecture is incorrect for $n = 8, 20$.

Although there is no perfect analogy between isoperimetric relationships and those true for solids in 3-space, we did use, judiciously, previous knowledge to explore volume-surface relationships in certain solids. We were able to amass some knowledge of rectangular prisms, spheres, cylinders, and cones.

15.2   PROBLEMS FOR CHAPTER 15

1.   a)   Measure several cylindrical cans and calculate the volume and surface area of each. Fill in a table with the diameter and the height, the surface area and the volume, and the ratio $\frac{S}{V}$. (Note that minimizing $\frac{S}{V}$ is the same problem as finding maximum $\frac{V}{S}$.)

d	h	S	V	$\dfrac{S}{V}$	Comments

b) Draw an arrow to the can of minimum ratio on your list. Is it absolute minimum?

c) Now consider the set of all cylindrical cans having a certain volume, V. Is there any connection between the minimum ratio and that of $\frac{d}{h}$?

d) Are there any two cans that have different dimensions and yet have the same ratio, $\frac{S}{V}$?

2. What is the best topless box you can make with a given amount of material? It must have the greatest volume for its surface.

3. What is the best topless box you can make for the money if the wood in the bottom costs twice that of the wood on the sides? That is, what is the shape that will hold the maximum volume?

4. Make up a problem of your own and show its solution.

1. a) The first uses $2 \cdot 16 + 2 \cdot 40 = 2 \cdot 96 = 192$ square units; the second box uses 184 square units.

    b) The second box. $\dfrac{160}{184} > \dfrac{160}{192}$.

2. $4 \times 4 \times 4$ is the box of maximum volume.

3. a) $S/2$ b) V^2 c) $(xy) \cdot (yz) \cdot (xz)$ must be maximized.

4. Out of this rectangle, we fashion some sort of prism of height h; $V = A_{base} \cdot h$. The sides enclose maximum volume when the base encloses maximum area; the base area is maximum when circular, forming a cylindrical container.

5. The right circular cylinder.

6. a) $\dfrac{V}{S} = 1.09$ for cylinder.

    $\dfrac{V}{S} = .455$ for cone.

    The cylinder is the better choice of container for the packaging material used.

    b) Yes, always, because $2h < 2s$, so $2h < r + 3s$ and $\left(\dfrac{V}{S}\right)_{cone} < \left(\dfrac{V}{S}\right)_{cylinder}$

7. Surprisingly, the regular octahedron does not contain maximum volume. One cannot generalize quite so lightly from two to three dimensions.

ANSWERS TO PROBLEMS

1. a) Each can has its own ratio of S to V.

    b) There is no can of absolute minimum ratio, $\dfrac{S}{V}$. You may see a pattern, however, when examining many cans of the same volume.

    c) Yes, the minimum ratio, $\dfrac{S}{V}$, for a certain volume occurs

    when $d = h$

    or $\dfrac{d}{h} = 1$.

    Here is how you find that the ratio is minimal when $d = h$.

    $V^2 = \pi r^4 h^2$.

    $2V^2 = \pi [2r^2 \cdot hr \cdot hr]$.

    $\dfrac{2V^2}{\pi} = 2r^2 \cdot hr \cdot hr$.

    $S = \pi [2r^2 + hr + hr]$.

    $\dfrac{S}{\pi} = 2r^2 + hr + hr$.

    $\dfrac{S}{\pi}$ is minimum for $\dfrac{2V^2}{\pi}$ when $2r^2 = hr = hr$

    or $2r = h$.

    Hence S is minimum for V^2 when $2r = h$, and thus S is minimum for V under these conditions.

    d) Yes, when the ratios, $\dfrac{r + h}{rh}$, are equal.

2. Half a cube because double the surface contains maximum volume when the box is in the shape of a cube.

3. The shape of a cube.

BIBLIOGRAPHY

Archimedes. *The Method.* Edited by T. L. Heath. New York: Dover Publications, Inc., 1912, pp. 10–11.

Bush, Grace A., and Young, John E. *Geometry for Elementary Teachers.* San Francisco: Holden-Day, Inc., 1971, pp. 224–56.

Courant, Richard, and Robbins, Herbert. "Plateau's Problem." *The World of Mathematics.* Vol. 2. Edited by James R. Newman. New York: Simon & Schuster, Inc., 1956, pp. 901–9.

———— *What is Mathematics?* New York: Oxford University Press, 1960, pp. 394–7.

Haldane, J. B. S. "On Being the Right Size." *The World of Mathematics.* Vol. 2. Edited by James R. Newman. New York: Simon & Schuster, Inc., 1956, pp. 952–7.

Polya, George. "Mathematical Discovery." *Induction and Analogy in Mathematics. Mathematics and Plausible Reasoning.* Vol. 1. Princeton, N.J.: Princeton University Press, 1954, pp. 155–8, 166.

A SUMMARY OF THE RELATIONSHIPS BETWEEN DIFFERENT GEOMETRIES STUDIED IN CHAPTERS 1–15

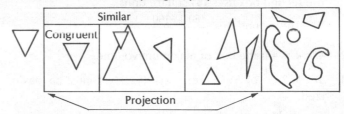

Something topologically equivalent:	any simple closed curve.
Something that is a projection (omitting one special projection):	any triangle.
Something that is similar:	any triangle having sides whose lengths are in the same ratio as the original and having equal respective angles.
Something that is congruent:	one triangle having exactly the same angles and lengths of line segments.

chapter 16

coordinate
geometry

MATERIALS NEEDED:

Several yardsticks and rulers

Chalkboard size compass and protractor

Several compasses and protractors

Dowell rod on a platform

In this chapter you will examine in detail several simple relationships between geometry and algebra, and learn to refer algebraically to some familiar curves, surfaces, and solids. The power of algebra had widened the applications of geometry and has allowed much of that study's growth. The study of geometry from that standpoint is called *analytical geometry*. Coordinatization of familiar shapes enables us to apply the power of algebra to description and to the solving of many types of problems. If you are interested, you can even see how the coordinatizing of a cone allows for ease in finding its volume. ⏀

> ⏀ Note to Prospective Teacher:
>
> You may wonder about the advisability of introducing algebra in grade school when the field of geometry is so rich. But you will no longer think this when you finish this chapter and begin to grasp the power of algebra both as a tool opening up far greater explorations in geometry and as an instrument widening the applications of geometry.

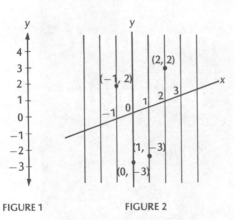

FIGURE 1   FIGURE 2

We can verbally describe certain sets of points— for instance a line on the paper parallel to the left edge and 5 inches to the right of that edge. But we need a system if we want to be able to indicate each point, each curve, each surface, and each solid. We choose to develop a system that is algebraic.

To identify a relationship between algebra and geometry, we must look at the basics of each discipline. In their beginnings, algebra deals with numbers and geometry deals with points; utilizing algebra in connection with geometry necessitates attaching a numerical name to each point. The process of naming points numerically is called *coordinatization*. Now to the groundwork for a coordinate type of system.

A point on a specific number line such as line y in Figure 1 is easily located; we locate $y = 2$ with a dot in Figure 1. But some expansion is needed in labeling a randomly chosen point on this page. If it is on the number line y we can specify it algebraically as a positive or negative number on that line, depending on where we put the zero and how big the units are. But if it is off that line, we need to give more information in describing its position. Consider the family of number lines parallel to our original number line y. Some of these can be seen in Figure 2, along with one intersecting line, x. The intersecting line, x, gives us a way of denoting the appropriate vertical number line in the family.

Two numbers can be used to locate a point on this page in this way: the first number locates the particular vertical line by counting along on line *x* from the intersection of *x* and *y*, the second number denotes distance up or down that vertical line, the zero point being the point of intersection with line *x*. In Figure 2, four points are labeled according to their location in this system.

When line *x* is perpendicular to the vertical lines and the same unit of distance is used on line *x* as on the vertical lines, we call our frame of reference a Cartesian coordinate system.

Another system of coordinatizing the plane involves considering a point as lying on one of a set of circles with the center at point 0. You can denote a point by indicating the circle on which it lies and then by indicating where it lies on that circle. To do this, specify which circle by stating a distance along a horizontal ray through 0, and then tell where the point is on that circle by stating angular degrees.

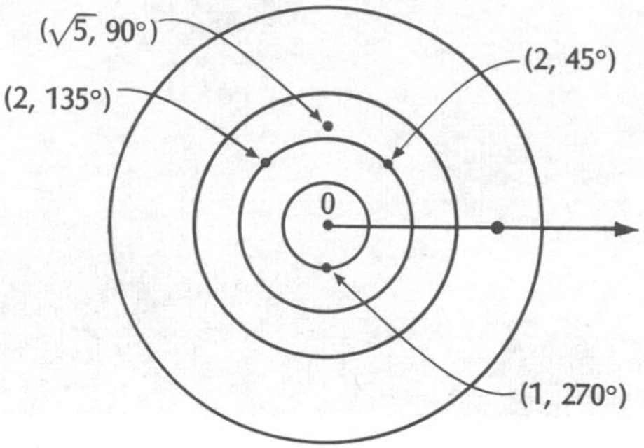

FIGURE 3

Figure 3 shows such a system with the points (2, 45°), ($\sqrt{5}$,90°), (2,135°), and (1,270°) marked on it to show you how to use a horizontal ray to the right from 0. We call these coordinates polar coordinates. (It is especially easy to write the equation of a circle in polar coordinates. The circle of radius 5 is simply written $r = 5$.)

[1]The navigator's latitude-longitude reference system is a coordinate system valuable on a sphere.

These ideas will allow you to invent other coordinate systems as you need them.[1] If possible, incorporate a preferred ingredient of any coordinate system, a one-to-one relationship between points in the plane and pairs of numbers. Compare the systems of Figures 2 and 3, and you will see that, in one of the systems, each pair of numbers denotes a separate point and that each point has just one corresponding number pair. Which figure demonstrates such a system? Sometimes the one-to-one relationship is so important that the reader's choice is restricted to such a system. Consider this property preferable if you need to invent any coordinate system.

EXERCISE 1:

Fill in the following table, then tell which figures portray a one-to-one property.

Figure	No. of points for one number pair	No. of pairs for one point
2		
3		

16.2 CARTESIAN COORDINATE SYSTEM IN TWO DIMENSIONS

A particular system of the type shown in Figure 2, but having a horizontal line for the reference line, is called a _Cartesian coordinate system_ after its originator, the Frenchman, René Descartes (1596–1650). Descartes reportedly was lying on his back on a cot watching a buzzing fly near the ceiling. Nothing might occur to us at a moment like this, but he was a mathematician and a philosopher, so it occurred to him that he could indicate the position of the fly by three numbers signifying perpendicular distances from the upper corner of the room. (The first two numbers locating a point on the ceiling, and the third number stating how many units below the ceiling the fly buzzed.) If this story is true, the idea of a coordinate system was born in the usual visual space of three dimensions. For ease in learning, 2–dimensional coordinatization is presented first even though historically it was seen as a special case of 3-dimensional geometry—the case where one coordinate is 0.

In the Cartesian coordinate system for 2 dimensions, the horizontal reference line is a real number line, referred to as the _x-axis_. Label a vertical line y and call it the _y-axis_, then refer to the intersection of the x-axis and the y-axis as the _origin_. A point in this system can be well-demonstrated by a fly sitting on a window with a corner of the frame serving as the origin. A set of points in the plane takes on numerical meaning when the particular coordinate system is specified; this means specifying units along both axes. The Cartesian coordinate system is well-defined if the respective axes are labelled and on each axis the numerals 0 and 1 are clearly marked.

◑ Note to Prospective Teacher:

Here is a good physical exercise using a corner of the room as the origin, (0, 0), with one of the walls denoted as x-axis.

a)   Stride off, as accurately as you can, the points (5,2) and $(\frac{1}{2}, \frac{7}{2})$, using 1 step to equal 1 unit.

b)   Compete in teams of two for speed in plotting, with a chalk and a yardstick, a list of five points on the floor. (Note that the winners demonstrate, in their efficient communication of duties, a knowledge of the importance of order in the number pairs.)

c)   Each student writes, on a piece of paper, the position of his desk to the nearest foot's accuracy. Use a pair of numbers in parentheses as you did in part a.

Note the graphs in Figures 4 and 5, their algebraic expressions, and the accompanying partial table of values indicating points that lie on each curve. It is instructional to fill in a few more entries in each table as you examine each graph.

EXERCISE 2:

Graph $x - y = 2$ for $-1 \le x \le 2$, (for x values between -1 and 2). This should be a line segment, a part of Figure 5.

Elementary exercises in the plane make use of the floor, the chalkboard, pegboards, and grid paper. These include plotting points, finding distances between points, and graphing all the points that fulfill a numerical relationship. Here are a few to set the mood and to trigger your creativity.

EXERCISE 3:

Visualize yourself and a friend competing against other teams in the race described in the note to the prospective teacher. ◑

What division of duties would you use if you had a corner of a room to yourselves and were handed a list beginning (5,2) and $(\frac{1}{2}, \frac{7}{2})$.

Sketch the floor in the corner with these two points.

EXERCISE 4:

Draw your own Cartesian coordinate system on a piece of plain paper.

FIGURE 4

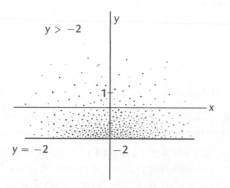

$y > -2$

y

1

x

$y = -2$

-2

Shading Denotes the Set of All Points
Described by $y > -2$.

	$y = -2$
x	y
5	-2
-1800	-2

FIGURE 5

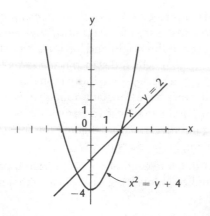

y

1

0 1

x

$x - y = 2$

$x^2 = y + 4$

-4

$x - y = 2$	
x	y
5	3
2	0

$x^2 = y + 4$	
x	y
0	$-y$

a)   Plot (0,8) and (0,−2); determine the distance between them.

b)   Plot and find the distance between (3,4) and (−9,4).

c)   Plot and find the distance between (0,8) and (3,4).

This last exercise might send you to the geoboard to affirm that you can stretch a rubber band joining any two points so that it loops around a third point and forms a triangle having two perpendicular line segments. Application of the <u>Pythagorean Theorem</u> then yields the exact distance. Check yourself against an example.

EXAMPLE:

Find the distance between (0,−2) and (−3,7).

Two-step solution:

Appropriate third vertex to make a right triangle: (0,7) or (−3,−2)

By Pythagorean Theorem, distance is $\sqrt{3^2 + 9^2}$
$= 3\sqrt{10}$

EXERCISE 5:

a)   Transfer to grid paper one of the pegboard right triangles that go through (0,−2) and (−3,7). Fill in the x and y-axes and the origin and label the correct lengths along the three sides of the triangle.

b)   Can a right triangle for which you know all three lengths be formed with the hypotenuse drawn between $(1/2, 7/2)$ and $(-5/2, 5)$? State why not or show your labelled successful attempt on grid paper.

c)   Develop a descriptive formula for finding the distance between $(0,0)$ and (a,b).

d)   Develop a descriptive formula for finding the distance between (x,y) and (a,b).

EXERCISE 6:

For each part of this exercise, set up a Cartesian coordinate system on half of a sheet of grid paper, graph, and label or shade in some way the set of all points such that

a) $x = 1; x < 1$.

b) $x + y = 1; x + y < 1$.

c) $x + y > 2$

d)   The number of gallons of gas consumed $= 1/20$ of the number of miles driven. (Customarily the horizontal axis is reserved for whichever variable appears the most independent—in this case, the miles driven.)

e)   The number of pounds $= 2.2$ times the number of kilograms an object weighs.

f) $F = 2A + 5$

g)   The temperature on Fahrenheit scale, F, is $32 + 9/5$ times number of degrees Centigrade.

EXERCISE 7:

A surface bounded by a square can be written as $\{(x,y): 0 \leq x \leq 1 \text{ and } 0 \leq y \leq 1\}$. (This notation

stands for the set of pairs (x,y), where x is between 0 and 1, and y is between 0 and 1.)

a)   Graph the above square surface and state its area in square units;

b)   Repeat part a for the surface described as $\{(x,y): -1/2 \leq x \leq 1/2 \text{ and } -1/2 \leq y \leq 1/2\}$.

EXERCISE 8:

a)   Graph *and label* 10 points of distance 1 unit from the origin and coplanar with the origin.

b)   Do you see a pattern in the position of these points with respect to the origin or to each other? Sketch lightly over the points with a pencil to indicate the pattern you see.

c)   What shape is formed by the set of all points *in space* that are 1 unit from the origin?

EXERCISE 9:

a)   Try finding a numerical relationship between x and y that is true for all points, (x,y), on a circle centered at the origin and having a radius of 5. It helps to state which distance is the same for all points on a circle, since there is an expression in x and y that has a constant numerical value for all points on the graph. If you are not grasping something intuitively, try elucidating the relationship by filling in the pairs below.

b)   Complete each of the six indicated pairs to symbolize points lying on a circle with a radius of five.

FIGURE 6

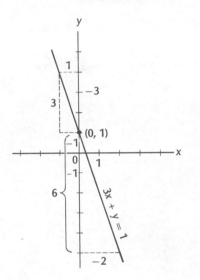

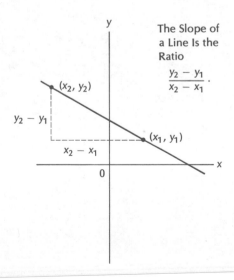

The Slope of a Line Is the Ratio
$$\frac{y_2 - y_1}{x_2 - x_1}.$$

(0,) (,0)

(3,) and (,4)

(−4,) (,−5)   List more if you need it.

Study of a line:

Note that the graph of $3x + y = 1$ is a line going through (0,1) and having a slope upward to the left. (See Figure 6a). This slope is -3, a number that represents the ratio of the number of units up to the number of units left, $\dfrac{3}{-1} = -3$. A horizontal line is said to have slope 0. A line slanting upward to the right has a positive slope—a positive ratio because the denominator is positive, representing units to the right. The slope of a line is defined as the ratio of vertical and horizontal distances between two points on the line—any two points since the ratio is always the same.

Notice that, in Figure 7 and henceforth, we use an equation as a convenient name for a line. As previously shown, the line $3x + y = 1$ has a slope of -3. You can see this slope showing in the rewritten equation that reads $y = -3x + 1$, where the slope is displayed as the coefficient of x.

To prove: $3x + y = -5$ is parallel to $3x + y = 1$. You can tell that these lines are parallel to each other because if the lines intersect in some point (x_1,y_1), then those numbers satisfy the equations for both lines.

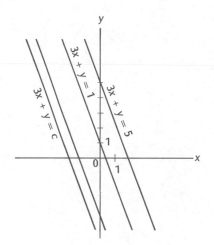

FIGURE 7   Notice That the Number Called c Indicates the Point at Which the Line $3x + y = c$ Crosses the y-axis.

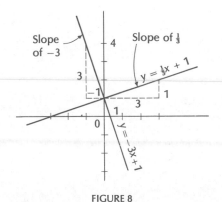

FIGURE 8

If (x_1,y_1) is a point on $3x + y = -5$,

then $3x_1 + y_1 = -5$.

If (x_1,y_1) also lies on $3x + y = 1$,

then $3x_1 + y_1 = 1$.

Both these statements cannot be true, since the same number cannot be both -5 and 1.

Hence, there is no point lying on both lines.

Note that the two parallel lines both had slopes of -3. Each line parallel to $3x + y = 1$ has slope of -3. The family of all lines parallel to $3x + y = 1$ can be described merely by specifying the slope. Hence, the family of all lines parallel to $3x + y = 1$ can be expressed by the equation, $3x + y = c$, where c is any real number.

The popular pair of lines, two perpendiculars, are related in their algebraic expressions. Examine $y = -3x + 1$ and the line perpendicular to it that passes through (0,1). The slope of the perpendicular is opposite in sign. What is its slope? (Notice the congruent triangles emphasized in Figure 8).

A line perpendicular to $y = -3x + 1$ is one having a slope of $1/3$, the negative reciprocal of the original. The set of all lines perpendicular to $y = -3x + 1$ is a set of parallel lines having a slope of $1/3$. Use previous knowledge of parallel lines to write the family of lines perpendicular to $y = -3x + 1$. The family of all such perpendiculars can be described by the general equation, $y = 1/3\,x + c$, where c is a real number. The perpendicular passing through (0,1) is described by the equation $y = 1/3\,x + 1$.

The line $y = 1/3x + 1$ passes through (0,1). The point (0,1) is on each line of the form $y = mx + 1$, since (0,1) makes each such equation true. As you would expect, the lines through (0,1) all have different slopes, varying as m is assigned different real numbers. Two of these lines are shown in Figure 8.

EXERCISE 10:

Find the slope of the line segment whose endpoints are (3,1) and (4,−6).

EXERCISE 11:

Express algebraically the family of all lines

a) parallel to $y − x = 4$.

b) perpendicular to $y − x = 4$.

c) through (5,6).

EXERCISE 12:

In each of four cases find the missing coordinate so that the line passes through the two points

a) and is horizontal $\begin{cases} (8,4) \text{ and } (0,\ ) \\ (a,b) \text{ and } (c,\ ) \end{cases}$

b) and is parallel to $\begin{cases} (1,1) \text{ and } (4,\ ) \\ (-3,-5) \text{ and } (4,\ ). \end{cases}$
    $3x − 2 = y.$

EXERCISE 13:

Find the fourth vertex of the rectangle

a) that has three vertices, (2,−2) (8,−2) (8,3).

b) that has three vertices, (3,−1) (7,2) (4,6)

EXERCISE 14:

Without plotting the points, prove that (0,2) (−3,−7) and (5,17) are collinear (lie on the same line).

EXERCISE 15:

The vertices of a triangle are (1,3) (4,4), and (−1,9). Use slopes to prove that the triangle is a right triangle.

EXERCISE 16:

The vertices of a quadrilateral are $A(0,0)$ $B(5,0)$ $C(8,4)$ and $D(3,4)$.

a) Show that the diagonals are perpendicular.

b) What special kind of quadrilateral is $ABCD$?

16.3 MORE THAT IS POSSIBLE IN TWO DIMENSIONS*

The following section will be a lecture on some of the connections between coordinate geometries and previous material. The star by the title of this section signifies that you need not demonstrate proficiency in this material; you can still gain a sense of direction and see displayed some of the usefulness of the subject. The bibliography at the end of the chapter lists texts, any one of which will serve as a good supplement and source of instructional material should you choose to spend a few more days on this chapter.

We have been looking at families of lines. There are other families of curves—families of circles, families of ellipses, etcetera. Concentric circles can be expressed easily in a coordinate system, for we lose none of the shape or size by imagining them all centered at some specific point such as the origin. In polar coordinates, the set of all concentric circles is described by the equation $r = c$, where c is any constant. In Cartesian coordinates, the same idea prevails except that the radius of the circle is expressed in terms of x and y. So, $\sqrt{x^2 + y^2} = c$ describes all concentric circles.

EXERCISE 17:

a)   Express, in Cartesian coordinates, the set of all circles that have their centers at the origin.

b)   Below this, express, in Cartesian coordinates, the set of all circles having their centers at (2,3). (Hint: First draw one such circle and express the distance from the center to any point on that circle.)

The tremendous power of coordinatizing becomes apparent when you first transfer a problem into algebraic language. Coordinatization is especially helpful for proving something about the slopes of lines or lengths of line segments. Problems of this nature include proofs that, for example, two sides

are of equal length and parallel (or perpendicular) or that a certain line segment is bisected by a line.

The next two exercises include two theorems for you to prove by assigning coordinates to vertices of a polygon and working with distances and slopes of line segments. These are proofs using algebraic means rather than geometric. The reasoning is like that used in Exercise 16. To prepare your thought for this method of proof, note the simple proof below.

THEOREM

The length of the diagonal of a rectangle is expressed as $\sqrt{a^2 + b^2}$, where a is the length of the rectangle and b is the height.

PROOF

1.   Any rectangle can be placed so that one of its vertices is at the origin and another vertex on the x-axis.

2.   Placing the rectangle in this way, we get other coordinates specified for any rectangle of length a and height b. (See Figure 9.)

3.   The diagonal of positive slope extends from (0,0) to (a,b), and thus has slope $\dfrac{b}{a}$ and length $\sqrt{a^2 + b^2}$ by the Pythagorean Theorem. A quick check will affirm that the other diagonal has the same length.

To coordinatize the vertices of a parallelogram,

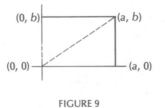

FIGURE 9

place it similarly with the lower left vertex at the origin. It then takes only a bit of thought to label the vertices so that we get a height of b and two parallel bases of length a. Convenient coordinatization is half the work in some problems. Now use these helps to try a few proofs on your own.

EXERCISE 18:

Prove (using algebraic means) that the diagonals of a rhombus are perpendicular. (A rhombus is a parallelogram having all sides of equal length.)

For the next proof, you have to be able to label coordinates of midpoints of line segments. This is easy when you recall the popular drawing for similar triangles and see that halfway up the hypotenuse implies half the horizontal and half the vertical distances. (Study Figure 9′)

EXERCISE 19:

Prove that the line segment connecting midpoints of two sides of a triangle has certain properties:

a)   that its length is half that of the third side of the triangle.

b)   that it is parallel to the third side.

You may remember the shape formed by the set of all points that were possible third vertices of is operimetric triangles formed on $\overline{AB}$, when the measure of $\overline{AB}$ was 6 meters and the total perimeter was 16 meters. Reexamine Exercise 10a of Chapter 13 and you will realize that the curve in Figure 10a

FIGURE 9′

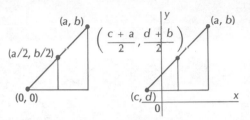

FIGURE 10

a) Complete This Partially Formed Ellipse by Using the Fact That the Sum of the Two Distances from A and B Is 10. Label Points C and D and Those Points Opposite Them.

below is described as the set of all points whose distances from A and B add up to 10 units. It is called an ellipse. See Figure 10a and b.

DEFINITION

An *ellipse* is the set of all points lying so that the sum of the distances from two fixed points is constant. The two fixed points are called *foci*.

The Greeks found it necessary to make a fantastic construction and superposition in three dimensions in order to find the foci of an ellipse. Coordinatizing makes it so much easier, and coordinatizing has made available the tool of algebraic calculation from which emerges a simple formula for finding the foci still faster. (You can find this formula in textbooks on analytical geometry.)

16.4   COORDINATIZATION IN THREE DIMENSIONS

The corner of a room as pictured in Figure 11a serves as a good model of the meeting of three perpendicular positive axes labeled as in Figure 11b. The fly in Figure 11b is at (3,1,2). This section is about points in space, coordinatization in three dimensions.

Make certain that you can locate points from reading ordered number triplets and can state approximate coordinates of things in your classroom when a corner is denoted as the origin. As a

Figure 10 Continued

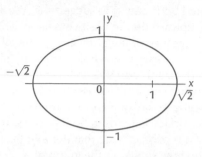

b) Use the Definition of an Ellipse to Find the Foci of This Ellipse. Label the Foci.

FIGURE 12

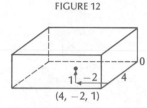

Locating a Point in the Room by Giving Three Numbers

FIGURE 11

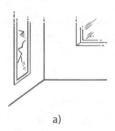

a)

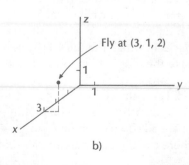

b)

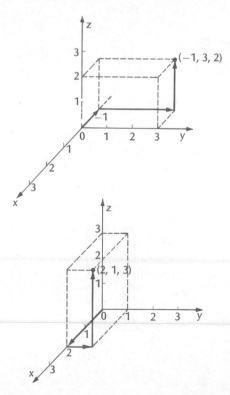

convention, we list first, the distance along the x-axis, then the distance parallel to the y-axis in the xy-plane, and lastly, the z-distance. If you were quick at locating points in the plane, you will find the adaptation simple. To find (3,1,2), just locate the point (3,1) on the floor (calling it the xy-plane), and then visualize the point moving directly upward z units. Hence (3,1,2) is directly above (3,1)—exactly 2 units above it. The point (3,1,−3) would be 3 units down into the room below. We usually consider the section of space seen in the room as the set of points where x, y and z are all positive. The locating of points in a room comes through experience. Note the paths used in plotting each point in the next three drawings. (Figure 12)

A less limited aid to exploring in three dimensions is a simple platform with a dowell rod sticking out of its center, like some of us tossed hoops over. One good use of this apparatus lies in considering the platform as the xy-plane and the dowell rod as the z-axis. Make sure you see that the part of space above the platform is the set of all points (x,y,z) where z > 0.

The dowell rod, handy for its moveability, has another advantage in that you can show the set of points in the four octants for which z > 0. (There are eight octants, the portions into which space is separated by three intersecting planes; in Cartesian coordinates, the planes are mutually perpendicular and are named the xy-plane, the xz-plane, and the yz-plane. Notice that when z > 0 you are above the xy-plane.)

Figure 4 showed the grapy of $y = -2$ in the plane. To denote that line in a 3-dimensional system, we would have to specify that both $y = 2$ and $z = 0$. (If we fail to specify z, we have more than a line.)

In the dowell rod model, we can lay a yardstick down parallel to the x-axis and two units "back of the dowell rod" (away from the observers) to model this line. If the corner of a room is used traditionally as that octant in which all variables are positive, we cannot show this particular line because one of the variables is negative.

EXERCISE 20:

Hold a yardstick near the corner of the room to indicate the line $x = 3$ and $z = 0$, and then move the yardstick slowly to a position where $x = 3$ and z is 4 units.

If z can be any real number, what is the set of all points such that $x = 3$? Describe the set verbally or sketch its graph on the coordinate system of Figure 11b.

We can use the idea of surface extended infinitely in learning how to coordinatize other surfaces besides planes. Consider an elliptical cylinder. The set of all points satisfying the equation $9x^2 + y^2 = 1$ is an ellipse in the plane where $z = 0$. (Figure 14a) Because the equation leaves z unrestricted in a 3-dimensional system, it denotes a surface suggested by a pile of congruent ellipses extending upward on the dowell rod as z becomes greater and greater. Figure 14b displays a graph of this sym-

FIGURE 13

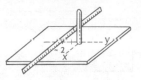

FIGURE 14

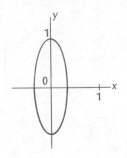

a) $9x^2 + y^2 = 1$ in the Plane.

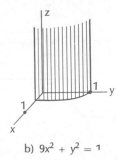

b) $9x^2 + y^2 = 1$

metrical surface as seen in one octant where the incompleted boundary is to suggest that the surface continues upward (in the positive z-direction) infinitely. (Because of the symmetry of the elliptical cylinder we can suggest a whole surface by picturing it in one octant. You should be able to infer the rest of the elliptical cross-sections in the four quadrants above the xy-plane and then see that this whole surface extends in both positive and negative z directions.) Note that this elliptical surface is somehow similar to the graph of $x = 3$ in space; they both are easily graphed by noting a cross-section and noting that z is unrestrained. Because z is unrestricted in the equation, we see, directly above and below each point in the xy-plane, a whole set of points for the infinite vertical line of z-values.

Note the graphs in Figure 15. Certain algebraic equations are easy to graph. If you wish to check your understanding of easy 3D graphing, skip to Exercises 22 and 23 at the end of the chapter.

We have been considering the set of all points that form a plane or some other surface when one variable is left unconstrained in the algebraic equation. It matters not what variable is unrestricted, we can sketch its axis in any position we wish. These surfaces in which the third variable is unrestricted are the easiest to graph and visualize, but by no means are they the only ones that can be graphed. For instance, $3x + y - z = 5$ is a plane, as you can affirm by careful plotting of points. (So are all surfaces of the form $ax + by + cz = k$, where a and b and c are not all zero).

FIGURE 15

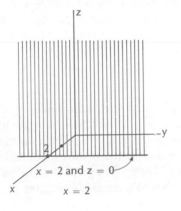

x = 2 and z = 0

x = 2

x = 2

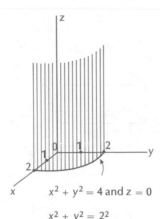

$x^2 + y^2 = 4$ and $z = 0$

$x^2 + y^2 = 2^2$

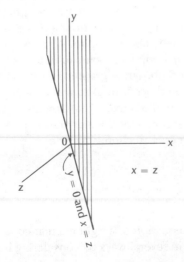

x = z

y = 0 and x = z

Synthetic geometry, the kind that proves from a postulate system such as Euclid's, contrasts unfavorably with analytical geometry when visualizing gets rough. Synthetic geometry gets harder to picture and visualize in 3D, but the part of analytical geometry that is not dependent upon visualization does not get significantly more difficult. This is because the algebraic aspects get only a little more complicated for three and more dimensions. This is the principal advantage of analytical geometry, that it substitutes for visualization formal manipulations such as finding points of intersection.

Many surfaces and solids discussed in geometry can be expressed algebraically. The equation of the solid is obvious from the equation of the bounding surface; for example, all points within the cylinder $x^2 + y^2 = 2$ are described by $x^2 + y^2 < 2$. Hence, awareness of a few other surfaces is sufficient to tie in much of synthetic geometry to the analytical method. Some other surfaces that are advantageous to consider algebraically are spheres, cones, and general surfaces of revolution.

The sphere is the set of all points at a certain distance from a point. Place the point at the origin, and we can express the sphere in several ways. The sphere of radius 2 is denoted by the equation $\rho = 2$ in still another coordinate system, spherical coordinates. Expressing the equation of a sphere of radius 2 is more difficult in Cartesian coordinates because we must relate the values of x, y, and z to the number 2. Before trying to find this equation, let us look more closely at the graph of a sphere centered at the origin of a Cartesian coordinate system.

EXERCISE 21:

a)   Sketch a sphere centered at the origin of a Cartesian coordinate system and having a radius of 2 units.

b)   Use a red pencil to trace the part of the sphere that intersects the xy-plane.

c)   State the equation in the xy-plane of the red tracing made in part *b*.

You no doubt noticed in this exercise that each cross-section of the sphere was a circle. In the xy-plane (where $z = 0$) lies the great circle described by the equation $x^2 + y^2 = 2^2$. Our equation should be such that when $z = 0$, $x^2 + y^2 = 4$. But how can we find an equation for a sphere when x, y, and z are all varying? It is not as hard as it sounds. We must find a way of expressing a distance between a point (x, y, z) and the origin, and then it is easy to express the sphere of radius 2 as the set of all points such that this distance from the origin is 2.

Examine the first and second pictures of the rectangular prism shown in Figure 16. The distance between any two points in space can be seen as the diagonal of a right rectangular prism drawn in around these two points. The diagonal of the rectangular prism at right is the hypotenuse of a right triangle that has the z distance as one side and the diagonal of a rectangle as the other. The diagonal of

FIGURE 16

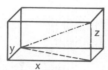

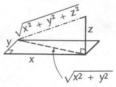

the base rectangle is $\sqrt{x^2 + y^2}$ because the x and y distances are solely involved. Then the square of cube's diagonal is of length $z^2 + (x^2 + y^2)$. So the distance between any two points is expressed as the square root of the sum of the squares of the x, y, and z distances. For example, the distance from (1,4,2) to (−2,3,−5) is $\sqrt{3^2 + 1^2 + 7^2}$. And the distance from (0,0,0) to (x,y,z) is $\sqrt{x^2 + y^2 + z^2}$.

So the sphere of radius 2 is the set of all points such that $\sqrt{x^2 + y^2 + z^2} = 2$. Note that the trace in the xy-plane (when $z = 0$) is indeed the circle $x^2 + y^2 = 4$. (Algebra confirms our expectation that we should again see a great circle when $x = 0$, and that we should see a circle of lesser radius when $x = 1$.)

As you remember, the sphere is a very symmetric surface. We chose to place the sphere where we did relative to our coordinate axes because of its symmetry about a point. When we notice symmetry about a point, it occurs to us to make that point be the origin. When graphing a right circular cylinder, it should occur to us to make the line of symmetry be one of our axes. That simplifies the picture—and it simplifies the algebraic expression, for then each cross-section is a circle.

The Cone and its Volume:*

The equation of a right circular cone also is simplified in several ways by considering it a surface of revolution about a line passing through the center of each of its circular cross-sections. Examine the cone in Figure 17, where we purposely place the cone so that the line $x = z$ is revolved about the z-axis. The radius of the circle at each altitude is equal to the x distance. Because $x = z$, this is also the z distance. So radius squared $= x^2 + y^2$ becomes $z^2 = x^2 + y^2$, the general equation of the cone in this position. Since a piece of a cone can be thought of as the outside of a spinning triangle, we can use our knowledge of triangles to make the equation for any cone. For the line $z = 2x$, the right circular cone formed has a radius of $\frac{1}{2}z$, and thus an equation of $(\frac{1}{2}z)^2 = x^2 + y^2$. Do you see how the algebra is simplified by advantageously positioning a figure in a coordinate system?

Another advantage of coordinatizing the cone and placing it about a line of symmetry is ease in finding its volume. Do you recall that in studying volume we summed little cylinders as an estimate of volume within a cone of finite height, getting successively better approximations as we decreased the heights of the little cylinders. Well, the fact that the line generating the side of the cone is expressible algebraically allows us to denote the heights and radii of the little cylinders no matter how many there might be. We can now denote the radius and height in general. Position the second cone in Figure 17 on its side so that it is formed by revolution about the x-axis. Let us assign it a height of 5 units to make the calculation more obvious. (See Figure 18.) For each x, a little cylinder has radius $\frac{1}{2}x$ and height $5/n$, where n is the number of

FIGURE 17

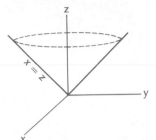

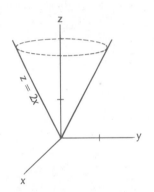

FIGURE 18

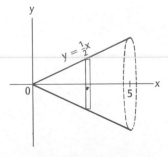

cylinders. It is with this coordinatization of a curve that calculus has the opportunity to express the exact volume within a cone as the limit of the sum of inner cones as the number of cylinders gets increasingly greater.

The volume of each slender cylinder is $\pi(\frac{1}{2}x)^2 \cdot 5/n$, where n is the number of cylinders. The mathematician asserts that $V = \int_0^5 \pi \left(\frac{x}{2}\right)^2 \cdot dx$, exact because of the theory of integration as it applies to this straight line (with this equation) revolving about the x-axis. The volume is said to be exactly "the integral from zero to five of $\pi \cdot (\frac{x}{2})^2$ differential x," which is $\frac{125\pi}{12}$ in case you are curious. If you have not yet been exposed to integral calculus, you may be surprised to see what a large proportion of the study of this field relies directly on geometry and coordinatization. The above brief comment on volume can serve as a bridge to another field—the finding of volume using the tool of calculus.

How we place a figure in its coordinate system determines its algebraic expression. The simplification of the algebra via clever placement cannot be overstressed. If there is plane symmetry in a figure, it behooves us to place the figure so that the plane of symmetry is a coordinate plane. If a figure is symmetric about the xz-plane, for instance, the appearance of (x,y,z) in the figure means that $(x,-y,z)$ is also in the figure. Thus, if y is replaced by $-y$ in an equation for the figure, the equation must still be the same. This means that y^2 and y^4, for example, may appear in the equation, but no odd powers of y. Point, line, and plane symmetry are each taken into account in placing a geometric figure. Careful placement of a geometric set of points in a coordinate system can lend insight and simplify the algebra and drawings. This savvy is acquired through experience, as is the ability to graph quickly. It is not the point of this chapter to give you that experience, but to provide a tie-in with synthetic geometry and give you sufficient direction for growth. Look to the following two exercises to give some practice in graphing, but realize that extra study in a book on analytical geometry is suggested if you desire much competence and speed in graphing, placing figures, and writing equations.

EXERCISE 22:

a)   Consider an equation in x and y representing a circle in the xy-plane; describe verbally what shape that equation represents in a 3-dimensional coordinate system.

b)   Sketch $x^2 + y^2 = 1$ in 3 dimensions.

EXERCISE 23:

Graph $x - y = 2$, employing a 3-dimensional coordinate system like that used in Figure 11b.

a)   Enter in the table below ten or more values and graph those ten points. Graph more, if they are needed to see a familiar shape.

b)   What is the name we give to this set of points?

c)   Graph $x - z = 2$ on a 3-dimensional coordinate system. What shape is the graph?

Values fitting $x - y = 2$.

x	y	z
3	1	18
3	1	−5
−2	−4	15

1.

Figure	No. of points for one number pair	No. of pairs for one point
2	1	1
3	1	infinite number

Figure 2 portrays a one-to-one property.

2.

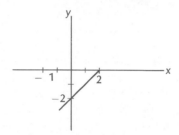

3.

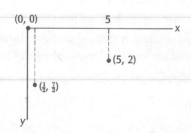

4.

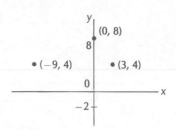

a) distance is 10.

b) distance is 12.

c) distance is 5.

5. b) Yes, a right triangle can always be formed.

c) $d = \sqrt{(a - 0)^2 + (b - 0)^2}$. d) $d = \sqrt{(a - x)^2 + (b - y)^2}$.

6. c)

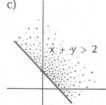

d)

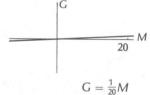

$G = \frac{1}{20}M$

7. b)

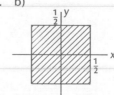

Area = 1^2

8. b) circle.

c) sphere.

9. a) 5 = distance of
    point from (0,0).

    The picture at right
    may help.

    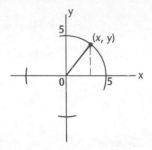

    b) (0,5)   (5,0)
        (3,±4) (±3,4)
        (−4,±3) (0,−5)

10. slope of −7.

    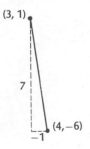

11. a) $y − x = k$, where k is any real number.
    b) $y + x = k$.
    c) $y − 6 = kx − 5k$.

12. a) (8,4) and (0,4).
        (a,b) and (c,b).
    b) will have slope of 3
        (be parallel to $3x − y = 2$).

    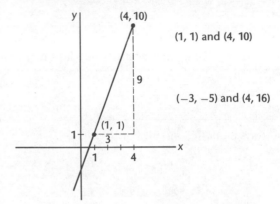

    (1, 1) and (4, 10)

    (−3, −5) and (4, 16)

13. a)

    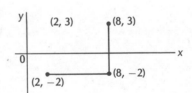

    b) (0,3).

14. Can be done either of several ways:

    1) Slopes between any two points are the same.

    2) If not collinear, a triangle is formed and the distances
    along two sides would exceed the length of a third side.
    If collinear, the distances add up exactly.

    By the first method, slope of $\overline{P_1 P_2} = \frac{1}{3} =$ slope of $\overline{P_2 P_3} =$
    slope of $\overline{P_1 P_3}$.

16.

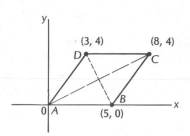

(3, 4) · (8, 4)

D

C

(5, 0)

0 | A · B

y

x

a) Proof that the slope of $\overline{AC}$ is the negative inverse of the slope of $\overline{BD}$. Slope of $\overline{AC} =$ $4/8 = \frac{1}{2}$. Slope of $\overline{BD} =$ $-4/2 = -2$. Thus the line segments are perpendicular.

b) a rhombus

17.

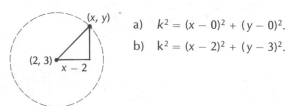

(x, y)

(2, 3)

x − 2

a) $k^2 = (x - 0)^2 + (y - 0)^2.$

b) $k^2 = (x - 2)^2 + (y - 3)^2.$

18. You are on the right track if you have labeled the four vertices and written down the slope of each diagonal.

19.

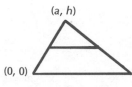

(a, h)

(0, 0)

Place triangle horizontally

One vertex is (0,0), and one of the mid-points has coordinates $\left(\dfrac{a}{2}, \dfrac{h}{2}\right)$ if the upper vertex has coordinates (a,h). Use what you know about similar triangles to continue labeling. You are on the right track if you label all five points so carefully that the midpoints are labeled in terms of the vertices. After labeling, compare lengths and compare slopes.

20. $x = 3$ is a plane in a 3-dimensional Cartesian coordinate system.

21. a)

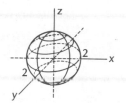

z

2

x

2

y

b)

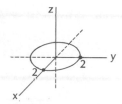

z

2

2

y

x

c) $x^2 + y^2 = 2^2$, in the 2-dimensional coordinate system.

22. a) $x^2 + y^2 = k^2$ is a cylinder of radius k in a 3-dimensional system.

b)

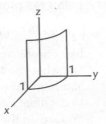

23. b) a plane.   c) $x - z = 2$ is a plane.

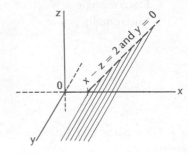

BIBLIOGRAPHY

Bush, Grace A., and Young, John E. *Geometry for Elementary Teachers*. San Francisco: Holden-Day, Inc., 1971, pp. 224–56.

Hutchinson, M. W. *Geometry: An Intuitive Approach*. Columbus, Ohio: Charles E. Merrill Publishing Co., 1972, pp. 259–302.

Kindle, Joseph H., Phd. *Theory and Problems of Plane and Solid Analytic Geometry*. New York: Schaum Publishing Co., Inc., 1950.

Love, Clyde E. *Elements of Analytic Geometry*. New York: The Macmillan Co., 1950, pp. 1–31, 150–4, 164–5, 176–7.

Oakley, C. O. *Analytic Geometry*. New York: Barnes & Noble Books, 1954.

Rosenthal, Evelyn B. *Understanding the New Mathematics*. Greenwich, Conn.: Fawcett Publications, Inc., 1966, pp. 111–47.

Schaaf, William L. *Basic Concepts of Elementary Mathematics*. New York: John Wiley & Sons, Inc., 1966, pp. 101–16.

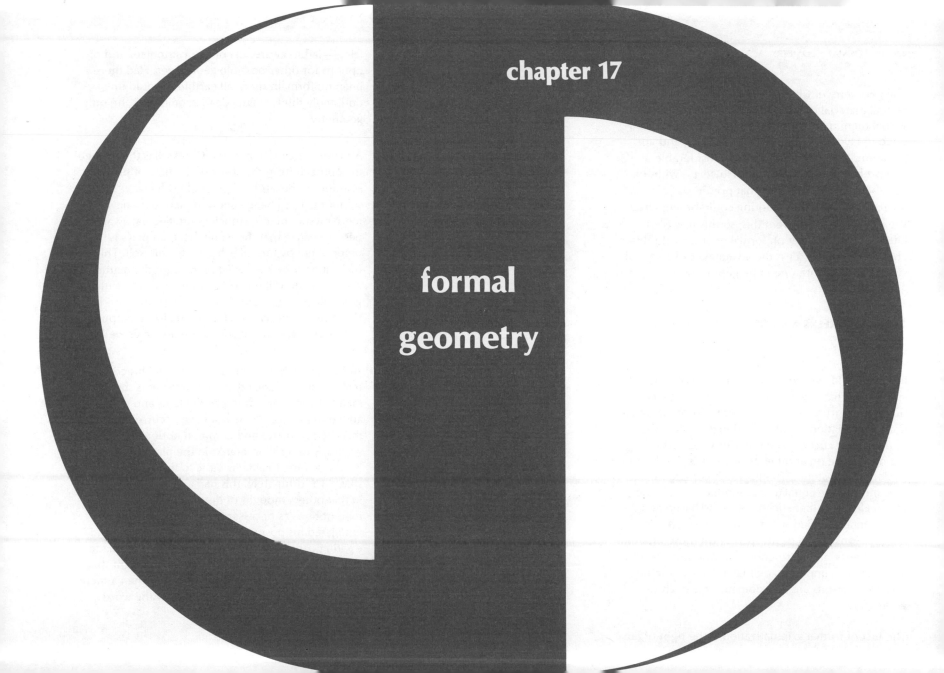

chapter 17

formal
geometry

The geometry of Chapters 1–15 involved you in the factual material of Euclidean geometry, although verbal memorization of these facts was de-emphasized in favor of intuitive understanding and application. You should now have considerable comprehension of these facts. However, without the ability to read any text that presents geometry formally, your further learning could be impaired. We would like to prevent this, so this text concludes with a chapter on formal geometry. In this chapter, we will discuss the advantages of a formal system and explore one finite geometry.

17.1 THE GREEKS AND US

In Chapter 10, we explored Euclidean congruence and found that any two sets of points can be considered congruent if one is a combination of rotational, translational and reflectional transformations of the other. Euclid listed this idea of superimposing one figure on another as one of the common notions in a system. "Things which coincide with one another are equal (congruent) to one another." In the *Elements* that Euclid wrote, he based geometry on a specified set of 23 definitions, 4 common notions, and 5 postulates. Upon these he built his geometry of constructions and theorems about measure. The importance of Euclid lies in the fact that his *Elements* are an axiomatic approach to geometry.

The fact of formal axiomatization—the tight organ-

izing—led to awareness of inconsistencies and of options for other possible geometries. Had there been no formalization, all earthfolk would undoubtedly think of Euclidean geometry as the only geometry.

A major contribution of the Greeks was the change in approach from the study of the how of problem solution to the study of principles (the why) of solutions. This change allowed man to go beyond mere mysticism into sufficient intellectual awareness to realize that the game that man plays is determined by the rules he sets for himself. This realization took over 2000 years to hit the mathematical world (till the 1820's when 2 non-Euclidean geometries were published); it is just now striking the general population that this realization applies to self-concept and to social interaction as well.

At the time when Euclid wrote, he and his contemporaries perceived as true statements the things man believed regarding parallel lines and equals added to equals. It may not have occurred to Euclid that the postulates and common notions he stated were any other than planks in the platform of reality, the geometrical facts basic to the geometry of space. Unfortunately, this had a stultifying effect in that others thought of these postulates and common notions as necessarily true. Until others investigated his postulate system further and found exciting consistent full-size geometries that made sense and yet were different, people thought that geometry was a model of reality rather than a mere man-made structure. The meaning of the word *geo-metry*, earth measure, indicates the belief of the Greeks that it was an empirical science rather than a body of assumptions and definitions and proven results (theorems).

Today, geometry is typically presented as a formal system where the student, with guidance, proves that certain theorems follow from certain other statements. Because of the preponderance of this presentation in texts, it is wise for you to have a little experience reading and exploring a formal geometry. Hopefully, you can also experience some of the joys of "getting it all together" and making brilliant deductions.[1] Reading a geometry presented formally is a quick and efficient way of covering material and of perceiving the geometry as a unified total in which can be seen interconnections and possible options. May this chapter help you to study, to "get it all together" and open to you another playground, the playground of formal geometry.

Euclidean geometry, which has a large and complicated formal structure, is poorly suited for our study because of its familiarity. We need a formal system that is unfamiliar so that the reader can better prevent entrance of unstated, and thus unwanted, concepts. Exploring an unfamiliar system allows the reader to build his own model of a geo-

metry in which all the postulates are true, skirting the assumptions of Euclidean geometry in which all lines are "straight" lines with an infinite number of points on them.

17.2 STAR GEOMETRY

The finite geometry presented here is called *Star Geometry* after the Star of David that inspired it and is one of its models. This geometry was created by a student working on her master's degree. A similar design could inspire any one of us to invent a geometry if we put our minds to it. The star geometry can be found in its entirety in her master's thesis.[2]

A geometry, like any abstract mathematical system, has certain basic elements:

1. Deductive reasoning and linguistic syntax.

2. A set of primitive, or undefined terms.

3. Postulates (a set of statements covering the undefined terms).

4. Theorems (statements concerning the undefined terms that can be deduced logically from the postulates).

We are relatively accustomed to deductive reasoning and language usage. (It's in our culture.) It is the undefined terms and the statements concern-

---

[1] Note to Prospective Teacher:
Deducing has a puzzle-like quality that warrants elementary school classes' playing with formal geometries on the level of James E. Major's "Rings and Strings," *The Arithmetic Teacher* (October 1966), pp. 457–60.

---

[2] Margaret M. Pratt, "Finite Geometries," (Master's thesis, San Jose State College, 1964), pp. 124–33.

ing them that get our directed attention in examining the star geometry.

Star Geometry

The *plane* consists of a set of *pointas* that are undefined, together with certain subsets of this set called *linas* and an undefined relationship between pointas and linas called *on* such that the following postulates are satisfied.

POSTULATE 1

There exists a lina on 4 pointas.

POSTULATE 2

Every pointa lies on exactly 2 linas.

POSTULATE 3

Any two distinct linas have, at most, 1 common pointa.

DEFINITION

Two linas that have no common pointa are said to be <u>parallel</u>.

POSTULATE 4

There is exactly one lina in the plane that is parallel to a given lina. (*P4* is a distinguishing characteristic of this geometry.)

The first question to occur to a geometer relates to

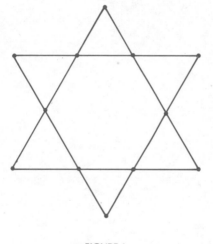

FIGURE 1

the consistency of this set of postulates, for that amounts to making sure that there is a model that will satisfy all the postulates. The postulates are consistent, for Figure 1 displays a model for which all the postulates hold true. Check to see that the star in Figure 1 with 12 pointas and 6 linas fulfills the postulates, preferably re-creating the model on scratch paper as you read through the postulates. Then take your eye off this model and answer the following questions on the basis of the postulates and what you can deduce from them.

EXERCISE 1:

Could the geometry have exactly 3 linas?

EXERCISE 2:

Must each lina have at least 2 pointas?

EXERCISE 3:

Prove to us that the geometry cannot have exactly 5 linas.

EXERCISE 4:*

Convince us that it is necessary to have more than 7 pointas.

In answering these questions, you find out some facts about the geometry resulting from this set of postulates. These facts are called theorems of the geometry. Here are eight theorems, some of which you may have already deduced in answering the above questions. It is necessary to your growth that you sketch all possible cases as you read a proof and that you write down something that completes a proof in each instance where the proof is left incomplete. Each finished proof will have *QED* (quid erat demonstratum) written at the end.

THEOREM 1

There are at least four distinct linas.

PROOF

By *P*1, there is a lina with 4 pointas, so there is at least one pointa, *A*. By *P*2, there are exactly two linas on *A*, called *p* and *q*.

Let *r* be a third lina parallel to *p*.

Lina *r* cannot be parallel to *q*, by *P*4.

So there is a fourth lina, *s*, which is parallel to *q*. *QED*.

THEOREM 2

Each lina has at least two pointas.

PROOF

By Theorem 1, there are four distinct linas. Since each lina is parallel to one lina (by *P*4), each must intersect two remaining linas.

These pointas of intersection must be distinct because *P*2 guarantees that each pointa lies on only 2 linas. *QED*.

COMMENT

Theorem 2 and its proof comprise the full answer to Exercise 2, as Theorem 1 answers Exercise 1. Though you may have visualized sketches corresponding to the proofs of Theorems 1 and 2, you will find it almost essential to construct a sketch as you study each part of Theorem 3.

EXERCISE 5:

Construct a sketch of each of the two cases handled in parts I and II as you study the proof of Theorem 3.

THEOREM 3

Each lina has exactly 4 pointas.

PROOF

Let *A* and *B* be two pointas on lina *p*, the lina referred to in *P*1 as having 4 pointas. Let *q* be a second lina with *m* pointas, and let *C* and *D* be two of these pointas. Let *AB* denote lina *p* and *CD* denote lina *q*.

Case I.   Assume *AB* and *CD* are parallel. Each pointa on *AB* lies on one other lina, which implies that *AB* intersects 4 linas. Similarly, *CD* intersects *m* linas. Moreover, each lina intersecting *AB* must intersect *CD* also, inferring from *P*4. No two of these linas can intersect either *AB* or *CD* in the same pointa, by *P*2. Hence, the 4 linas intersecting *AB* must intersect *CD* in 4 distinct pointas, and *m* must be at least as large as 4. Suppose *m* is greater than 4. Then there is at least one lina intersecting CD that does not intersect AB, contrary to P4. Hence $m = 4$.

Case II. Now assume AB and CD intersect in a pointa, E. Let FG denote the lina parallel to AB, and let HI be parallel to CD. A result of part I tells us that FG has 4 pointas as does its parallel, AB, and that HI has m pointas along with its parallel, CD. One of the linas crossing CD is AB, and another is FG (since FG is not the parallel). By P2, there are $m - 2$ linas on CD other than those which also lie on AB or FG.
Each of these linas intersect both AB and FG in one of their pointas, of which 2 remain after CD and HI intersection pointas are counted.
These two unnamed linas intersect CD in 2 distinct pointas. So $m = 4$. QED.

THEOREM 4

There are exactly $4 + 2$ linas.

PROOF

Let p be a lina in the geometry. By theorem 3, it has 4 pointas, as does every lina. There are 4 linas intersecting p. Its parallel excepted, all linas in the plane must intersect p. Counting p and its parallel, there must be $4 + 2$ linas altogether. QED.

DEFINITION

Two pointas that do not lie together on any lina are said to be _parallel_ to each other.

THEOREM 5

Each pointa has at least one parallel pointa.

PROOF

Let AB and CG be any two parallel linas, and suppose A is a pointa without a parallel.

.

.

.

THEOREM 6

There are exactly $q = \dfrac{4(6)}{2}$ pointas in the plane.

THEOREM 7

Each pointa has exactly $p = q - 2(4) + 1$ parallel pointas.

Postulate P1 need not have used the particular number 4. As different numbers replace the 4 in that postulate, a whole family of geometries result. Let us denote as P1' the general statement of P1.

POSTULATE P1'

There exists a lina on n pointas, where n is a positive integer.

Everything we have done so far can be considered the result of substituting 4 for n in P1'. The effect of various substitutions is first noticed in the models, then in the theorems.

EXERCISE 6:

The model for $n = 4$ is the star of David shown in Figure 1. Make a model, if possible, of the geometry implied by $P1'$ through $P4$, where $n = 2$ in $P1'$.

EXERCISE 7*:

Make a model, if possible, of the geometry implied by $P1'$ through $P4$, where $n = 3$ in $P1'$.

EXERCISE 8:

Can the n referred to in $P1'$ be any positive integer, or are there some values of n for which no model exists?

Now we can take a look at the last of our eight theorems of star geometry.

THEOREM 8

The integer n mentioned in $P1'$ is even.

PROOF

Suppose n is odd.

Then the number of linas in the plane, $n + 2$, is odd by Theorem 4.

It follows that there is some lina that either has no parallel or has more than one parallel.

As this is impossible, $n + 2$ must be even, and so must n. *QED.*

EXERCISE 9*:

Make a model that holds for the set of postulates, $P1'$ through $P4$, when $n = 6$.

Some of these eight theorems may need adjustment if you wish to see the set of theorems that follow from using the general postulate, $P1'$. Your intuition should tell you that $P1'$ will not change the first two theorems. The third and fourth theorems need to be checked to see if they are true when the integer 4 is replaced by n. (First check the proofs to see if they work when n is substituted for 4.) Theorems 5 through 7 could be similarly checked and reworded for the general case.

You have seen and played a game by the rules (postulates). Does it seem possible to get the same model and the same theorems if you alter one of the postulates?

EXERCISE 10*:

If you replace $P2$ with "Every pointa lies on exactly three linas," what model do you get for $P1'$ through $P4$, where $n = 2$ in $P1'$?

EXERCISE 11*:

Show that the game (the set of theorems) is changed in some way by the substitution of the $P2$ mentioned in Exercise 10 for the old.

Changing the rules changes the game. The postul-

ates are the rules. New geometries result whenever
a rule is changed. We call a geometry a game because there is some change in the game for each
change in the rules—like playing monopoly with a
creative youngster. Man, learning the facility to
make a change and felicitously explore its effects,
receives considerable understanding of an axiomatic system. To "give it the light touch" in classrooms, many axiom systems are introduced with
strange words replacing "point" and "line"—like
"oog" and "limp." Another approach is reinterpretation of points and lines to objects, for example
using persons for points and committees for lines.
Substitute into the postulates the words "person"
and "committee" and examine this reinterpretation
of the postulate set for star geometry; it reminds
one of the set a real-life rules we must sometimes
observe.

EXERCISE 12:

Write out the star geometry postulates, $P1'$, $P2$, $P3$,
and $P4$, using the words "committee" and "person"
for line and point, respectively." Remember,
"parallel" needs to be expressed in relevant vocabulary.

1. No.

2. Yes.

3. If it had 5 linas, then one lina is not parallel to any.
   This contradicts P4, so there cannot be 5 linas.

4.* By P1, there are 4 pointas on a lina, l1.

   By P2, there are 4 linas intersecting l1 at the 4 pointas.

   Since none of them are parallel to l1, there is another lina l2 so that l2 ∥ l1.

   There are 4 pointas of intersection of l2 and these other linas.

   Total of 8 pointas of intersection, at least.

5. Theorem 3. I.

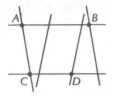

   Theorem 3. II.

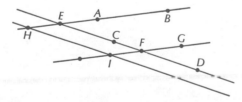

6.

7.* Impossible. The picture below begs for a parallel to l3. The addition of a fourth lina crossing l1, would force a fourth pointa on l1, by P2. Hence, it is impossible to find a lina on 3 pointas.

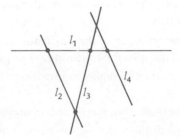

8. n cannot be an odd integer. There are geometries for each positive integer. If you can prove that, you show excellent understanding of this formal geometry.

9.* A model

10.*

This model works for n = 2 in P1′ and the new P2.

11.* The P2 substituted in the postulate system influences the theorems. As a start, Theorem 1 remains the same, for 3 linas on each pointa makes a system of l1 and two linas through a pointa on l1, a parallel to l1, and still more.

The reasoning of Theorem 2 is affected by the change in the second postulate.

Theorem 3 is difficult to read through, but is most likely still true.

Theorem 4 changes because the third statement of the proof is not true.

12. Such a reinterpretation of the postulate set for the star geometry would give the following set of postulates:

P1′.   There exists a committee consisting of n persons.

P2.   Every person is on exactly two committees.

P3.   Any two distinct committees have at most one person (member) in common.

P4.   There is exactly one committee that has no person in common with a given committee.

Bennett, A. A. "Modular Geometry." *The American Mathematical Monthly*. Vol. 27 (1920), pp. 357–61.

Courant, Richard, and Robbins, Herbert. *What Is Mathematics?* New York: Oxford University Press, 1960, pp. 214–27. (sophisticated, but gives models of hyperbolic and other non-Euclidean geometries)

Cundy, H. Martin. "25 Point Geometry." *Mathematical Gazette*. Vol. 36 (1952), pp. 158–66. (brief but interesting through p. 160. It takes work to affirm all he says.)

Eves, Howard. *A Survey of Geometry*. Vol. 1. Boston: Allyn & Bacon, Inc., 1963, p. 428. (a bit erudite, but good on history)

Exner, Robert M., and Rosskoph, Myron F. *Logic in Elementary Mathematics*. New York: McGraw-Hill Book Co., 1959, pp. 100–25. (good, uses "oog," for example, to avoid narrow interpretations of line and point)

Freund, John. *A Modern Introduction to Mathematics*. Englewood Cliffs, N.J.: Prentice-Hall, Inc., 1956, pp. 243–51. ("right on," and good exercises, A 6 point geometry)

Garstens, Helen L., and Jackson, Stanley B. *Mathematics for Elementary School Teachers*. New York: The Macmillan Co., 1967, p. 114

Jones, Burton W. "Miniature Geometries." *The Mathematics Teacher* (February 1959), pp. 66–71. (good clear motivated 7 point geometry including discussion of duality.)

Kasner, Edward, and Newman, James. *Mathematics and the Imagination*. New York: Simon & Schuster, Inc., 1963, pp. 114–5, 135–40.

Kenyon, Anne. *Modern Elementary Mathematics*. Englewood Cliffs, N.J.: Prentice-Hall, Inc., 1969, pp. 164–6.

Major, James E. "Rings and Strings." *The Arithmetic Teacher* (October, 1966), pp. 457–60.

Montague, H. F., and Montgomery, M. D. *The Significance of Mathematics*. Columbus, Ohio: Charles E. Merrill Publishing Co., 1963, Chapter 1. (patterns in geometry)

Pratt, Margaret M. "Finite Geometries." Master's thesis, San Jose State College, 1964, pp. 124–33.

Richardson, Moses. *Fundamentals of Mathematics*. New York: The Macmillan Co., 1958, pp. 440–58.

Ringenberg, Lawrence A. *College Geometry*. New York: John Wiley & Sons, Inc., 1968, Chapters 2 and 7.

Rosenthal, Evelyn B. *Understanding the New Mathematics*. Greenwich, Conn.: Fawcett Publications, Inc., 1966, pp. 217–27, 202.

Rosskopf, M. F. "Modern Emphasis in the Teaching of Geometry." *The Mathematics Teacher*. Vol. 40 (1957) pp. 272–9. (In discussing axiomatic systems, he gives [pp. 276–9] a 4-line geometry of 5 axioms and discusses consistency, independence, completeness, categoricalness, and the role of models.)

Schaaf, William L. *Basic Concepts of Elementary Mathematics*. New York: John Wiley & Sons, Inc., 1966, pp. 117–27.

Sharp, Evelyn, *A Parent's Guide to Mathematics*. New York: Simon & Schuster, Inc., Pocket Books, Inc., 1964, pp. 213–7.

Stabler, E. T. *An Introduction to Mathematical Thought*. Reading, Mass.: Addison-Wesley Publishing Co., 1953, pp. 139–48. (Same 6 point geometry Freund used. Duality explained well. Pages 146–8; consistency)

Stubblefield, Beauregard. *An Intuitive Approach to Elementary Geometry*. Monterey, Calif.: Brooks/Cole Publishing Co., 1969, pp. 27–31.

Ward, Morgan, and Hardgrove, C. E. *Modern Elementary Mathematics*. Reading, Mass.: Addison-Wesley Publishing Co., 1966.

Witter, G. E. *Mathematics: The Study of Axiom Systems*. New York: Blaisdell Publishing Co., 1964, pp. 168–223.

Wolfe, H. E. *Introduction to Non-Euclidean Geometry*. New York: Holt, Rinehart & Winston, Inc., 1945, pp. 46–47. (History of thought before 1824)

Young, John W. *Lectures on Fundamental Concepts of Algebra and Geometry*. New York: The Macmillan Co., 1911, Chapters 1–5, 13, 14, and 17. (good lectures)

appendix

SETS AND ACCOMPANYING VOCABULARY

First of all, what is a set? A _set_ can be thought of as some definite collection of objects. We use the word properly when we refer to a set of dishes or a set of tools. The objects can be symbols like letters of the alphabet or invisible entities like sets of points in space. The objects in the collection are called _elements_ of the set. If x is in the set, then x is an element of the set. Because points in space are the objects of geometry, there will be a tendency to use sets of points for examples in this appendix.

We could speak of all geometrical configurations as subsets of the set of points in space. Examples of subsets of 3-dimensional space are the set of points on a certain line segment, a certain equilateral triangle of 3-inch sides, a plane, a sphere, and a solid cube. Properly defined, set A _is a subset of_ set B (written A ⊆ B) if each element of A is also an element of B. (Note that the definition allows that B is one of its own subsets.) Among the subsets of a line are any point, line segment, or ray that lie on the line—and the line itself. Subsets of a cube include any face, edge, vertex, or plane section of the cube. The statement that a square is one plane section of a cube relates the first set to the second in the relation "is a subset of."

Another special relation frequently referred to is "disjoint." Two sets are _disjoint_ if they have no elements in common. Two parallel lines are disjoint and the line along the front edge of a chalk tray is disjoint from the plane of the blackboard to which the chalk tray is affixed. See Figure 1 for further examples.

FIGURE 1

Four Disjoint Sets of Points

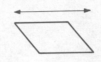

The _intersection_ of two sets is the set of elements they have in common. The intersection of two highways has been rightly named. It is the shaded portion of Figure 2. The set formed by the intersection of two lines has either one point in it (Figure 3a) or it is empty (Figure 3b), depending on the relative position of the two lines.

The set formed by the intersection of two parallel lines is an example of an _empty set_ or a _null set_, mathematicians' names for a set having no elements. According to zoological report, another example of an empty set might be the set of all giraffes that can fly.

Allowing for the possibility of a null set, intersection is always defined. Otherwise, the notation for the intersection of A and B would not always have meaning. Any sentence about A intersect B would be meaningless when A and B are disjoint. With the null set (symbolized by ϕ), the statement "sets A and B are disjoint" can be written "the intersection of A and B = ϕ." The use of the symbol ∩ for intersection makes the previous sentence even faster to write like this, "$A \cap B = \phi$."

Let's summarize the notation even though the symbol ∪ has not been discussed.

Sets:	English Terminology
ϕ	null set
$A \cap B$	A intersect B
$A \cup B$	A union B

Relations between
  sets:

$A \cap B = \phi.$ A and B are disjoint.

$A = B.$ A and B are the same set.

$A \subseteq B.$ A is a subset of B.

Figure 4 shows each of the options to consider when $A \cap B$ is mentioned. Examples include two rays intersecting in a line segment

$\overrightarrow{PQ} \cap \overrightarrow{QP} = \overline{PQ}$

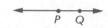

and two line segments intersecting in a point.

$\overline{ST} \cap \overline{QR} = m$

Important examples of the intersection of two sets of points are seen in the intersection of a plane and another surface. A plane intersecting a sphere is a circle or a point or ϕ; the intersection of a plane and a cube can be any of these sets: ϕ, one point, a line segment, a square region, a triangle, a quadrilateral, a pentagon, or a hexagon. These intersections are also called "slices" or "plane sections."

Two lines in space intersect in a limited number of possibilities. If L_1 and L_2 are lines in space.

Mathematical descriptions	Accompanying English
$L_1 \cap L_2 \neq \phi$	They intersect in one point or they are coincident lines.

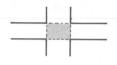

FIGURE 2

Intersection of Two Highways

FIGURE 3

The Two Possibilities of Sets Seen When Two Lines Intersect

FIGURE 4

$L_1 \cap L_2 = \phi$ and L_1 and L_2 lie in the same plane.   They are known as parallel lines.

$L_1 \cap L_2 = \phi$ and L_1 and L_2 are not in the same plane.   They are known as skew lines.

EXERCISE 1:

Observing Figure 5, describe and/or sketch all possible intersections of the following point sets.

a) $\overleftrightarrow{ab}$ and $\overleftrightarrow{cd}$

b) $\overleftrightarrow{ab}$ and $\overline{cb}$

c) $\overleftrightarrow{cd}$ and $\overline{ct}$

d) $\overleftrightarrow{cd}$ and $\overrightarrow{cd}$

e) $\overleftrightarrow{ab}$ and the half-plane above $\overleftrightarrow{cd}$

f) $\overline{ad}$ and $\overline{cb}$

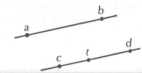

FIGURE 5

EXERCISE 2:

Draw a figure to illustrate each of the following statements:

a) $\overleftrightarrow{pq} \cap \overleftrightarrow{rs} = \{t\}$

b) $\overline{ef} \cap \overline{pq} = \overline{eq}$

c) $\overrightarrow{pq} \cap \overline{ab} = \overline{pa}$

EXERCISE 3:

Make three sentences using the verb, ⊆. Arrange them so that your three sentences are true for the figure you drew in Exercise 2c.

EXERCISE 4:

Draw in one plane two different angles whose intersection is

a)   empty

b)   four points

c)   three points

d)   a ray

The _union_ of two sets is the set of elements in one or the other (or both) of the sets. $A \cup B$ (A union B) is the set of all elements that lie in A or in B or in both. Note that the symbol for union is suggestive of a bowl upturned to include all; the idea of a bowl should help you differentiate between union and other symbols.

Notice also that the last diagram in Figure 6 is portraying $B \subseteq A$. In that situation you could refer to the rest of A as the _complement of B in A_ (symbolized as B′).

If $B \subseteq A$, then $A = B \cup B'$

B and its complement make up the whole set, A. Using the idea or the word complement often helps. To answer the question, "What is the set of points not outside a cube?", you can think of the complement in 3-space and reply, "The set of

points on the surface of the cube union those interior to the cube." The complement in the plane of a pair of lines is the set of points in the plane that are on neither of the lines.

Although in geometrical descriptions we can get out of using the word "union" often, we must also realize that use of this word makes some definitions especially easy. For instance, an angle can be said to be the union of two rays that have a common endpoint. A triangle is the union of the three line segments connecting three points that do not lie on the same line. And so forth for all polygons.

EXERCISE 5:

Let $\overrightarrow{OA} \cup \overrightarrow{OB}$ form one angle, called $\angle AOB$.

Let $\overrightarrow{OC} \cup \overrightarrow{OD}$ form another, called $\angle COD$.

The two angles lie in the same plane yet intersect in only one point, O.

a)   Sketch the union of the two angles.

b)   Describe in words the complement of that which you sketched.

EXERCISE 6:

Sketch or describe the first thing that occurs to you when you are given these "unions." (They are not all unique answers.)

a)   Union of a heart shaped patch and an arrow through it.

b)   Cube ∪ points inside the cube.

c)   Union of 2 circles having one point in common.

d)   Union of 6 identical square patches.

FIGURE 6

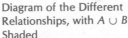

Diagram of the Different Relationships, with $A \cup B$ Shaded

There is a meaning to the intersection and union of three or more sets. Parentheses are helpful. In reading the set $(X \cap Y) \cap Z$, one finds the exact set denoted inside the parenthesis (in this case, $X \cap Y$), and only then does one perform the next operation outside the parenthesis (in this case, parenthesis intersect Z).

Following this procedure with the sets

$$X = \{1,2,4\} \qquad Y = \{1,2,6\} \qquad Z = \{2,4,6\}$$

you will find that $(X \cap Y) \cap Z = \{1,2\} \cap \{2,4,6\} = \{2\}$. Find $X \cap (Y \cap Z)$ for the same sets X, Y, and Z.

EXERCISE 7:

a) For X, Y, and Z related as shown in the diagram below, shade in the appropriate set beneath each of the identical pictures.

 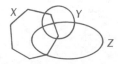

Shade $(X \cap Y) \cap Z$ Shade $X \cap (Y \cap Z)$

b) Does it seem likely that $(A \cap B) \cap C$ is generally the same set as $A \cap (B \cap C)$?

There are many diagrams showing the possible relationships of three sets, X, Y, and Z. You may find it very useful to draw as many as you can. These diagrams can be used to quickly check the relationship between $(X \cup Y) \cup Z$ and $X \cup (Y \cup Z)$. Doing this will convince you that parentheses are not needed for clarification when talking about the union of 3 sets or the intersection of 3 sets.

What you see as an associative property of intersection and an associative property of union may not extend to mixtures of the two set operations. Use the following exercises to start you on your way to seeing that the parentheses are necessary for clarification when union and intersection are mixed.

EXERCISE 8:

a) Find $S \cap (I \cup N)$ and $(S \cap I) \cup N$ for the sets,
$S = \{e,a,r,t,h,s\} \qquad I = \{h,e,a,d\} \qquad N = \{t,r,i,p,\}$

b) Does it seem likely that $A \cap (B \cup C) = (A \cap B) \cup C$ in general?

c) Show that the following theorem holds true for these sets, S, I, and N.
Theorem: $A \cap (B \cup C) = (A \cap B) \cup (A \cap C)$.

EXERCISE 9:

a) Find $(X \cup Y) \cap Z$ and $X \cup (Y \cap Z)$ for the same sets as used in the text. The sets are
$X = \{1,2,4\}$
$Y = \{1,2,6\}$
$Z = \{2,4,6\}$

b) Does it seem likely that $(X \cup Y) \cap Z = X \cup (Y \cap Z)$ generally?

c) Show that the following theorem holds true for the sets, X, Y, and Z.
Theorem: $A \cup (B \cap C) = (A \cup B) \cap (A \cup C)$.

ANSWERS TO APPENDIX EXERCISES

1. a) ϕ
   d) $\overrightarrow{cd}$

   b) The point b
   e) $\overleftrightarrow{ab}$

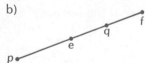

   f) The point m

   c) $\overline{ct}$

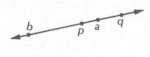

2. a)

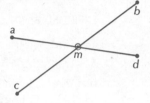

   c)

   b)

3. $\overline{pa} \subseteq \overrightarrow{pq}.$
   $\overline{pa} \subseteq \overrightarrow{ab}.$
   $\overrightarrow{bq} \subseteq \overleftrightarrow{bq}.$

4. a)

   b)

5. a)

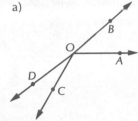

   b) The complement in the plane is the set of points other than those on any of the four rays.

6. a) Valentine.

   b) solid cube.

   c)

   d) $\Big\{$ a piece of rug, 2-by-3
   or
   $\Big\{$ a cube.

7. a)

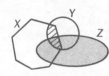

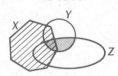

   $(X \cap Y) \cap Z = X \cap (Y \cap Z)$

   b) Yes, it is.

8. a) $S \cap (I \cup N) = \{e,a,r,t,h\}$
   $(S \cap I) \cup N = \{h,e,a,t,r,i,p\}$

   b) No

   c) $S \cap (I \cup N) = \{e,a,r,t,h\} = \{h,e,a\} \cup \{r,t\} = (S \cap I) \cup (S \cap N)$

9. a) $(X \cup Y) \cap Z = \{1,2,4,6\} \cap Z = \{2,4,6\}.$
   $X \cup (Y \cap Z) = \{1,2,4\} \cup \{2,6\} = \{1,2,4,6\}.$

   b) No.

   c) $X \cup (Y \cap Z) = \{1,2,4,6\} = \{1,2,4,6\} \cap \{1,2,4,6\}$
   $(X \cup Y) \cap (X \cup Z)$